"This book offers extremely useful guidance to business professionals and at the same time, translates very well to people in all walks of life. An excellent and impactful read!"

–Joe Juergensen, P.E., Muller Engineering Company

-

"John brings a client's perspective to engineering marketing. I would highly recommend this book as a required read for engineers transitioning into the marketing side of the business."

– David Center P.E., CFM. – AECOM

-

"John Burke is a respected professional in the engineering field and clearly understands the importance of relationships in our industry. I wish this subject was talked about more frequently – particularly in college – so more engineers would realize that relationship building is just as important as technical capability when it comes to winning work."

– Shea Thomas, P.E.,
Urban Drainage and Flood Control District

"Who knew engineers needed to have people skills? Engineering schools teach technical and problem solving skills, but there is very little focus on how to sell ideas, which is essential in the real world. This entertaining read offers guidance on what key qualities clients are looking for and then solid practical examples of how to personally develop these characteristics to become a successful leader in the engineering field. This book should be given as a graduation present to all engineering students."

– Laura Kroeger, P.E.,
Urban Drainage and Flood Control District

-

"Candid advice illustrated by interesting anecdotes to show how your approach to relationships can set you apart. A must read for all engineers!"

– Dave Skuodas,
Urban Drainage and Flood Control District

-

"Required read for anyone in the public sector dealing with consultant selection"

– Justin Werdel,
Colorado Department of Transportation

-

"Read this book. If taken seriously, you will absolutely challenge yourself to change the way you select design teams for your projects and deliver winning proposals."

– Steve Miller, Arapahoe County, Colorado

"John Burke provides inspirational guidance to professional development and success, winning people and projects. The thoughtful advice from a Client's perspective will help every Professional, Consultant and Salesperson understand how to better present themselves to others. Mr. Burke provides numerous life lessons and experiences to improve professional relationships. The book provides enjoyable stories of insightful direction for growth and personal development."

–Robert Krehbiel, Matrix Design Group.

-

"*Request for Personality* is a revealing and outstanding examination of the human mind, and what compels all people to make decisions. John returns the human component to project management and engineering to remind us how we all make decisions at our core. This is a must-read reference for anyone working in a technical industry."

– Brian Varrella, P.E., CASFM Chair

R.F.P.

REQUEST

FOR

Personality

WIN PEOPLE, WIN PROJECTS

JOHN D. BURKE

R.F.P. Request For Personality Copyright © 2014 by John D. Burke

While the techniques suggested in this book may have worked for some individuals and companies, this in no way guarantees these techniques will work for you. We do hope however the ideas presented within will assist you in winning more projects. Any representations in this book regarding consultant selection processes are solely for educational purposes and do not represent the bid selection or procurement process by any entity.

Dedication

This book is dedicated to the love of my life
– my best friend –
my wife, Julie, whose own journey in life
has made me a better man.

And to Halle, Alex and Olivia,
our three amazing kids who have blessed our lives
with a ton of smiles,
laughter, and wonderful memories
to last a lifetime.

Acknowledgments

First and foremost I want to thank my wife, Julie. It was her passion to write *Fire Station Baby*, the story to adopting our first child that inspired me to write this book. She is a wife, mother, author, sister, friend, and her strength and perseverance through struggling with chronic health is an inspiration to thousands.

Thank you Halle, Alex and Olivia – you are the three best kids any dad could ever want to share this life with. You are destined for greatness – always live your lives with passion.

Of course, I also want to thank my mom who always believed in me and planted seeds of greatness. To my brother David, who might be the only one in the world that shares my same humor – we've had many laughs together over the years. And to my brother Brian, who at 36 years old died suddenly of a heart attack last November, 2013 – I'll never forget you. You taught me to live each day to the fullest because you never know when it will be your last.

I also want to thank the most intelligent men I know that have mentored me in my engineering career: Pat Sorenson, Dan Giroux, Tom Giroux, Dave Jones, Mike Applegate, Dave

Downing, Dave Loseman and Steve Baumann. I've learned so much from each one of you and will be forever grateful for your friendship and making engineering fun, if that's possible.

A special thanks to Shea Thomas and Laura Kroeger with the Urban Drainage and Flood Control District for believing in the theme of this book so much that they have given me countless opportunities to speak and train at a number of events. Now you can just give people this book.

Thank you, Doug Schmidt, for editing and helping me navigate the path to publishing this book. Thank you, John Miller, author of *QBQ!*, for encouraging me during our brief time over coffee. Your message of personal accountability is one that is vitally important to everyone.

And finally, thank you to the many wonderful people that have poured their lives into such wonderful organizations as the American Council of Engineering Companies (ACEC), American Public Works Association (APWA), American Society of Civil Engineers (ASCE), Colorado Association of Stormwater and Floodplain Managers (CASFM) and the Colorado Stormwater Council (CSC). Each of these organizations has given me opportunities to learn and grow in my own personal journey to developing a winning personality.

Table of Contents

Introduction

Have you ever wondered why you didn't win a project when you were well qualified and ready to take on the work?

All things being equal it may have been something to do with the personality traits of your team.

You may already have these characteristics and can always further develop them – but it's possible that they're just not showing up in your proposal.

We're going to show you how to fix that.

-

"ALL THINGS BEING EQUAL, PEOPLE WILL WORK WITH PEOPLE THEY LIKE; ALL THINGS NOT BEING EQUAL, THEY STILL WILL."
— JOHN MAXWELL

-

This book was written from the perspective of those who must select design teams for their projects—the results of which may have career-long implications for decision-makers. When so much is at stake, those decisions must be made well.

As an example, I was recently asked to be part of the selection committee for a new ten million dollar high school. I'm part of the facilities committee for my kid's elementary school and they would someday attend this high school, so I wanted to make sure we had a great design-build team selected to do this work.

Dozens of hours were spent prequalifying the teams for this project and our job was to find the right one. We took a bit of an interesting approach where we interviewed the three architectural finalists and then the three contractors independently. At the end of the interviews, we put together the architect and contractor based on the value each would bring to the project.

Each company had one hour to make their best presentation and do their best to demonstrate why we should hire them. All the companies were pre-qualified and had long resumes of similar projects which were all outrageously successful of course.

Most of the architects with whom we spoke had worked with the contractors on one or more jobs – and they all said they'd happily work with whatever contractor we selected. When the contractors were asked about their prior working relationships with the architects we were interviewing, it felt a bit like a love fest – all of them gushing about how enjoyable it was to work with the architectural firms we were considering.

Having worked on dozens of similar projects, it was somewhat amusing to watch some of these contractors lie like that and talk about how much they loved working with the architects. Every contractor, at one time or another, comes to the end of their rope with the architect's idealistic concept of "creating a sense of destination" and incorporating Feng Shui concepts that simply cannot be constructed in the real world. At any given moment in time during construction the contractor is often just moments from burying the lead architect into the concrete foundation.

Nonetheless, our goal was to thin slice through their presentations and select which architect/contractor team that we'd hire for this work. The process was a little like speed dating, but in this case the contestants didn't see each other. The committee would select the winning design-build team based on their separate presentations—which, for us, was a strategic matchmaking process.

At the end of the architect's presentations, we had a lively debate on which one we all thought would bring the best ideas to the table, and would provide high quality at a good value. What made the difference? The architect we selected had done an exceptional job of *focusing on what was important to us.* Their proposal showed us conceptual renderings of our school. They had taken the time to use our school's logo and placed a draft building footprint on the proposed ten acre site that we were in the process of purchasing.

Their focus was entirely on us.

In contrast, the other two teams showed us renderings of projects they had completed over the past couple years. One focused on being cheap, the other focused on being functional. But neither school was our school. They were basically saying, "Look how well we have performed in the past, how wonderful we can be." In contrast, the winning team conveyed the idea, "You have an incredible vision here – look how wonderful YOU are."

◆❯ TAKE AWAY: FOCUS YOUR PRESENTATION ON YOUR CLIENT'S NEEDS, WANTS, AND DESIRES—NOT YOUR OWN.

If you're in any kind of sales position, this is a critical concept. People will always care more about themselves than you do, and when they find someone who agrees with their relatively-high view of themselves, they'll be drawn to hire you first.

That said, it was relatively easy to choose the architect as they had completely set themselves apart from the competition by focusing on the potential client instead of their resume.

The contractor selection process was a little more difficult. Let's face it, many contractors are good at building things, but throw them into a room and ask them to make a PowerPoint presentation is a little like asking them to throw on a tutu and do a number from Don Quixote.

Many times, it just ain't very pretty.

The bottom line is that all three contractors would bring value to the project. Each contractor had supplied a list of past projects, endorsement letters, standardized costs including overhead, profit and a well-honed statement of how excited they would be to work on this project.

After all the presentations were finished, the facilities committee came back together to review the proposals and determine which contractor to select.

After much debate, it came down to one thing: which contractor had *the best personality* to work with the architect we had chosen.

That was it – chemistry. No triple bottom line analysis. No phone calls to the references listed. We simply asked ourselves, "Can you see 'Bob' working effectively with 'Joe' through the design-build process?"

This is just one of many examples from my experience where the ability to work effectively with others became the sole basis for determining who to hire. All things being equal, the individual personalities of the project managers were by far the most important factor.

So, the question we all need to ask is this: *What's a winning personality when it comes to choosing a project team?* More importantly, if you were on the presenting end of things, how do you develop this winning personality that's going to get you and your team hired for most projects you choose to go after?

Or, if you're the one doing the hiring, what's going to attract the best teams to work for you?

I've always been fascinated by people who just seem to have a charismatic personality that's backed up by intelligence, character, and a strong desire to help clients achieve their goals. I'm talking about the best coaches, teachers, pastors, project managers and supervisors—mainly that one person you can always count on to find out what's important to you, and then give you wise counsel to head in that direction.

Working as a professional engineer for over twenty years now, I've seen teams that work well together and teams that don't. Without fail, the most productive and enjoyable teams are ones where the individual personalities of the group work together like a well-oiled machine.

But this kind of synergy rarely happens by accident.

Over the years I've kept a list of the qualities of teams that seem to tell me they've got what it takes. Some characteristics have been more important than others, depending on the situation. But, without fail, every time I think of how a team has made a positive impact or moves a project seamlessly through the process the strengths they bring can be found in these seven qualities:

1) Contagious Passion
2) Focused Vision
3) Connection speed
4) Positive Attitude

5) Unexpected Generosity
6) Fearless Integrity
7) Kaizen

These seven qualities are essential for winning just about any Request For Proposal (RFP) that comes out. And, they will be the reason why you'll get called back for future RFP's.

We're going to examine each of these seven qualities and provide specific tips on how to apply each of these qualities into your daily life in the workplace. At first, you will probably fail more often than you succeed. Over time, however, you will not only see positive change, but far more responsive clients.

The goal is to become the go-to person on any project – public or private, small or large. No matter what you do, you'll have the proposal-winning traits that clients are going to want on their teams.

chapter _1_

CONTAGIOUS PASSION

ROBERT WAS BORN in Daytona Beach Florida in 1942. He was a normal kid raised during World War II. He dropped out of school in the ninth grade and enlisted in the Air Force. He spent the next twenty years of his military career yelling and screaming at the enlisted men as he ascended to the rank of Master Sergeant. He always thought to himself that if he retired from the military, he would never yell or scream at anyone again.

Then he was stationed in Elmendorf Air Force Base in Anchorage Alaska. You can imagine how a kid from the beaches of Florida was awestruck by the majesty of the mountain ranges of Alaska. One day, Robert saw a TV show with Bill Alexander painting using a wet oil brush technique. He became entranced by this painter.

Robert was so interested in this oil painting technique that he contacted Bill and signed up for lessons. After a few months of disciplined practice he decided to start selling some of his work. Soon, he was making more money selling his paintings than he did in the military. So he retired from the Air Force and focused his time and energy painting and selling his artwork.

I was introduced to Bob in the mid-1980s the same way that many of you were. I happened to be in front of the television on one of those lazy Saturday afternoons flipping through all eight channels. I also became entranced in the way he swiftly moved the paint brush effortlessly across the canvas and created a majestic painting in just thirty minutes.

As he was painting, Bob would throw in these simple yet warming statements like "Let's just put a happy little tree over here, everyone needs a friend." Or, "We don't make mistakes; we just have happy accidents."

Bob Ross[1], with his soft spoken demeanor and iconic afro, captured the attention of millions of people. Every now and then, he'd also say something that brought a little sarcasm into his work. One of my favorites is, "Oh, if you have never been to Alaska, go there while it is still wild. My favorite uncle, Uncle Sam, asked me if I wanted to go there. He said if you don't go, you're going to jail. That is how Uncle Sam asks nicely."

Bob's passion and talent for painting drew us in. Many times after watching an episode I wanted to bust out to the nearest art supply shop, buy a canvas and some oil paints and learn how to paint like Bob.

Like Bob, each of us needs to exude contagious passion for what we do. To win that project, passion is a key element that doesn't cost us anything and will reward us more than just building a resume.

Think about it. Don't you just love being around people who are passionate about what they do?

For me, I immediately think of my kids. There have been many times when I got home from work and they all run up to me talking so excitedly that I can hardly understand what they were trying to say. But they are so enthusiastic that I couldn't help but be excited along with them.

It is a person's passion for life and what they do that makes them exciting to be around.

❯❯ CONTAGIOUS PASSION IS THE FIRST INGREDIENT FOR THE PERSONALITY THAT CONSISTENTLY WINS PROJECTS.

After watching an episode of "The Joy of Painting" with Bob Ross, anyone could see that his enthusiasm was contagious. It wasn't his wittiness or his amazing skill; we've all seen hundreds of paintings that look very similar to his work, rather, his passion that drew us into his world. This simple man, skilled in his technique and masterful at telling stories, has become an icon in American history.

Here are some lessons we can learn from Bob on being passionate for what you do:

1. Love doing it, even if you weren't getting paid.
2. Be good at what you do.
3. Enjoy teaching others.
4. Become a student and learn new things to teach.

At this point you may be thinking that you absolutely hate your job –that it's slowly sucking the life and passion out of you like a tick on an elephant's backside. *You just don't know what to do.*

Here's a suggestion — first, write down all the things you like about your job.

Ok, that may not have taken very long.

Now think about the things that you actually enjoy doing, perhaps even apart from your job description. Maybe it's mentoring and coaching other employees. Maybe it's leading a team discussion and listening to creative input. Perhaps it's implementing a process or procedure that streamlined operations. Maybe it's watching the construction of a project for which you made a significant contribution. Maybe it's about being able to solve complex problems.

List these things out, and consider how to apply them to different areas in your work life.

This exercise makes me think of John Miller, author of *QBQ! The Question Behind the Question.*[2] I first met John when he was giving a speech at my company and discussed the topic of personal responsibility. I was immediately drawn to his perspective and found it refreshing to get around someone who was willing to lay down some truth about what's at the root of most issues in the workplace. John shared many examples that day from other companies that he'd worked with over the years. His examples weren't very different from the ones my company faced on a daily basis.

I started following John's blog at QBQ.com. Since John lives only a few miles from me I asked to have coffee with him one morning and inquire about his inspiration for QBQ, his blog, books, and interviews on the topic of personal accountability.

He explained that while he was selling training programs for his company one of the things that hit him was how his clients seem to be asking "the wrong questions."

"The wrong questions" seemed epidemic when John would ask his clients what kind of training would best improve their company culture. John continually heard things like, "If we could only make Ted understand what he's doing wrong." Or, "Why did the sales team commit to an unrealistic deadline without checking with the manufacturing division?"

These types of questions are so commonplace we don't even recognize the fallacy in asking them. Play it out – what's the answer to these types of questions? Will they produce productive data that will lead to quality solutions? Or are they just deflective responses that further divide an organizational culture by assigning blame?

John's passion for getting to the right questions is what drove him to write a 299 page manuscript on the subject. He paired that down to 115 concise pages that became the version of *QBQ!* that sold over one-million copies!

I love hanging around people with this kind of passion for helping others.

Back to my question about what you're passionate about. I don't necessarily believe that passion for your job is the right answer. I believe you can find something that you're passionate about no matter what job or industry you're in. In fact, I don't think your ability to be passionate has anything to do with the type of work you do. You may be surprised to find that your deepest passion may come from your interaction with other people.

Do you love to help, teach, or lead others? Do you love to inspire and have fun with others? One of the greatest gifts we can give to others is to smile or laugh at the funny stories they tell us. How many times do we start the week off by telling each other interesting stories about the events of our weekend? These common interactions can lead to deeply satisfying, relationship-building opportunities.

In case you're having a hard time relating to what it looks like to be passionate in your job, let's take a look at what the opposite might look like. Here it is – The DMV (Department of Motor Vehicles). I've never seen anything that creates a passion-killing atmosphere as the "take a number" process.

Many of the employees seem to be cold and indifferent. The room is filled with anxious sixteen-year-olds hoping they'll pass their driving test, and their parents worried that they will. And then there's a host of other people praying they didn't forget some obscure document that would send them to the back of line, or back home, just to wait another two hours to be "served."

-

"THE QUALITY OF A PERSON'S LIFE IS IN DIRECT PROPORTION TO THEIR COMMITMENT TO EXCELLENCE, REGARDLESS OF THEIR CHOSEN FIELD OF ENDEAVOR."
– VINCE LOMBARDI, LEGENDARY FOOTBALL COACH

-

Think about your day-to-day operations at work. Weren't you a little more passionate for your job and your field of endeavor when you first started? At the same time, didn't you feel more energetic and interested in getting to know other people?

I've found that when people have enough respect for who they are they'll treat everything they do with an internal drive for excellence. In contrast, someone who lacks self-appreciation will never do his or her best work. It's almost as though they set an expectation of failure that never ceases to let them down.

Contagious passion is the key element that is often communicated subconsciously to other people around you. Passion is usually caught, not taught. Think about Steve Jobs the founder of Apple, Inc. His passion for making a better operating system and creating intuitive products has generated one of the most successful transformational companies ever to exist.

One of my favorite quotes from Steve Jobs is this:
"People with passion can change the world for the better."[3]

I believe that—only because I've seen it happen over and over again.

Another great example of someone whose personal conviction was greater than himself was brought to life in the movie "Amazing Grace." The story shares the true accounts of English statesman William Wilberforce.[4] For more than sixteen years he fought for the abolition of slave trade in the British Colonies. He risked his own wealth and well-being to fight for the Africans who were horrifically treated during this dark period.

Every year, beginning in 1791, William brought forth a new bill for consideration on the floor of the English parliament that would abolish the atrocities of slave trade. Unfortunately, with the amount of money at stake, estimated at around 80% of Great Britain's foreign income, his repeated bills were met with violent opposition.

Finally, after over sixteen long years of introducing bills just to see them get shot down, he was able to get an apparently benign piece of legislation passed in the House of Commons that effectively stopped slave trade immediately.

How long will you fight for the things that are important to you? Can the people around you see the depth of your passion even through the challenging times?

Find the things that you are passionate about. What are some things that you do that energize you? What brings an instant smile to your face when you think about it? What do you do that brings a smile to the face of others? What would you do even if you didn't get paid? What's something that you're naturally gifted at doing? What do others appreciate most about you?

Think about some of these questions and write down a few ideas. Then pursue them with contagious passion. Live every day with enthusiasm and purpose and see what starts to happen in your life.

There is something inexplicably transformational that happens when you start living with this quality.

When you live with passion, the rest of the world takes notice. You will begin to earn the respect and admiration of others and the next time you make a presentation in front of a prospective client, they will pick up on this passion directly or in a subliminal way. In either case, this is the "secret sauce" that will bring you more wins than losses.

So live with passion and watch opportunities come into your life like never before.

chapter 2

FOCUSED VISION

-

"THIS IS ONE SMALL STEP FOR A MAN; ONE GIANT LEAP FOR MANKIND."
– NEIL ARMSTRONG

-

THIRTEEN WORDS WERE SPOKEN 238,900 miles from the surface of the earth in 1969. These faint, crackling words traveled at 186,200 miles per second to receivers back on earth and then rebroadcast for the entire world to hear approximately 1.28 seconds after he spoke them.

Incredible.

To this day, when I look heavenward and try hard to think about the possibility of walking on the moon, I'm blown away that there have been people from this planet who have actually accomplished this.

The prelude to that historic event began many years earlier on May 25, 1961 when President Kennedy gave a speech before a joint session of the United States Congress.

What John F. Kennedy said before Congress was more than just ask them to fund the space program like never before. He asked them to take the huge risk and step with him and the thousands of engineers and scientists working diligently behind the scenes who told the president this feat was not only possible, but could be a hinge point in the history of the world as they knew it.

The Cold War was heating up as the Soviet Union was advancing Communism on one side, and the United States along with Allied countries tried to keep the Soviets at bay without provoking nuclear war.

During this time, another form of competition was in full swing – The Space Race. The utilization of outer space for counter intelligence and weapons deployment made advancing space exploration an absolute necessity. The delivery of this speech before Congress was timely as the Soviet cosmonaut Yuri Gagarin became the first person to orbit Earth in April, 1961. Shortly after that on May 5, 1961, American astronaut

Alan Shepard became the first American to travel into outer space.

The winner of the space race would set themselves apart from the rest of the world. The technology used to achieve that goal would change the way we live. It would be too costly not to join the race—and to win.

In his speech, President Kennedy said[1]:

> "Finally, if we are to win the battle that is now going on around the world between freedom and tyranny, the dramatic achievements in space which occurred in recent weeks should have made clear to us all, as did the Sputnik in 1957, the impact of this adventure on the minds of men everywhere, who are attempting to make a determination of which road they should take. Since early in my term, our efforts in space have been under review. With the advice of the Vice President, who is Chairman of the National Space Council, we have examined where we are strong and where we are not, where we may succeed and where we may not.
>
> Now it is time to take longer strides … time for a great new American enterprise … time for this nation to take a clearly leading role in space achievement, which in many ways may hold the key to our future on earth. I believe we possess all the resources and talents necessary.

But the facts of the matter are that we have never made the national decisions or marshaled the national resources required for such leadership. We have never specified long-range goals on an urgent time schedule, or managed our resources and our time so as to insure their fulfillment.

Recognizing the head start obtained by the Soviets with their large rocket engines, which gives them many months of lead-time, and recognizing the likelihood that they will exploit this lead for some time to come in still more impressive successes, we nevertheless are required to make new efforts on our own.

For while we cannot guarantee that we shall one day be first, we can guarantee that any failure to make this effort will make us last. We take an additional risk by making it in full view of the world, but as shown by the feat of astronaut Shepard, this very risk enhances our stature when we are successful. But this is not merely a race. Space is open to us now; and our eagerness to share its meaning is not governed by the efforts of others. We go into space because whatever mankind must undertake, free men must fully share.

I therefore ask the Congress, above and beyond the increases I have earlier requested for space activities, to provide the funds which are needed to meet the following national goals: First, I believe that this nation should commit itself to achieving the goal, before this decade is out, of landing a man on the moon and returning him safely to the earth. No single space project in this period will be more impressive to mankind, or more important for the long-range exploration of space; and none will be so difficult or expensive to accomplish.

We propose to accelerate the development of the appropriate lunar space craft. We propose to develop alternate liquid and solid fuel boosters, much larger than any now being developed, until certain which is superior. We propose additional funds for other engine development and for unmanned explorations—explorations which are particularly important for one purpose which this nation will never overlook: the survival of the man who first makes this daring flight. But in a very real sense, it will not be one man going to the moon-if we make this judgment affirmatively, it will be an *entire nation.* For all of us must work to put him there.

Secondly, an additional twenty-three million dollars, together with seven million dollars already available, will accelerate development of the Rover nuclear rocket. This gives promise of some day providing a means for even more exciting and ambitious exploration of space, perhaps beyond the moon, perhaps to the very end of the solar system itself. Third, an additional fifty million dollars will make the most of our present leadership, by accelerating the use of space satellites for world-wide communications. Fourth, an additional seventy-five million dollars—of which fifty-three million dollars is for the Weather Bureau—will help give us at the earliest possible time a satellite system for world-wide weather observation. Let it be clear—and this is a judgment which the Members of the Congress must finally make—let it be clear that I am asking the Congress and the country to accept a firm commitment to a new course of action, a course which will last for many years and carry very heavy costs: 531 million dollars in fiscal '62—an estimated seven to nine billion dollars additional over the next five years. If we are to go only half way, or reduce our sights in the face of difficulty, in my judgment it would be better not to go at all.

Now this is a choice which this country must make, and I am confident that under the leadership of the Space Committees of the Congress, and the Appropriating Committees, that you will consider the matter carefully. It is a most important decision that we make as a nation. But all of you have lived through the last four years and have seen the significance of space and the adventures in space, and no one can predict with certainty what the ultimate meaning will be of mastery of space. I believe we should go to the moon.

But I think every citizen of this country as well as the Members of the Congress should consider the matter carefully in making their judgment, to which we have given attention over many weeks and months, because it is a heavy burden, and there is no sense in agreeing or desiring that the United States take an affirmative position in outer space, unless we are prepared to do the work and bear the burdens to make it successful.

If we are not, we should decide today and this year. This decision demands a major national commitment of scientific and technical manpower, material and facilities, and the possibility of their diversion from other important activities where they are already thinly spread. It means a degree of dedication, organization and discipline which have not always characterized our research and development efforts. It means we cannot afford undue work stoppages, inflated costs of material or talent, wasteful interagency rivalries, or a high turnover of key personnel.

New objectives and new money cannot solve these problems. They could in fact, aggravate them further-unless every scientist, every engineer, every serviceman, every technician, contractor, and civil servant gives his personal pledge that this nation will move forward, with the full speed of freedom, in the exciting adventure of space."[1]

Let's take a look at some of the key things that Kennedy did in that particular speech:

1. **He set huge expectations** with great enthusiasm, but in a humble way. He made a strong declaration of the goal and the specific amounts of funding that would be necessary to accomplish the goal. However, even though he was the Commander-in-Chief, he was savvy enough to ask Congress to take action and humbly state his position.

2. **He gave credit to others** and empowered them. He mentions that vice president Lyndon B. Johnson was the chairman of the National Space Council. He also mentions how Captain Alan Shepard was the first American to travel into space.

3. **He identified the competition.** The Soviet Union was advancing Communism in Eastern Europe and the threat of nuclear war was closer than it had ever been in history. The Cuban Missile Crisis occurred in October of 1962.

4. **He appealed to a higher cause.** People support projects but they become ambassadors when you give them a great vision. Kennedy clearly drew a line in the

sand making the goal of landing a man on the moon not just an engineering feat, but a battle between good and evil. He opens his statement to Congress on this subject with words like "If we are to win the battle that is now going on in the world between freedom and tyranny."

5. **He clearly stated the risks of inaction.** The only way to assure defeat is to never attempt victory. "For while we cannot guarantee that we shall one day be first, we can guarantee that any failure to make this effort will make us last."

In these few words spoken to Congress, President Kennedy cast a vision so extraordinary and literally out of this world that it immediately created a buzz across a patriotic nation that wanted to succeed and beat the Soviet Union at the same time.

Think about the job you have. How can you explain it in such a way that gets others interested in what you do? Say you just heard that a company was looking to give away a few million dollars to a variety of organizations in your community and all you had to do was make a fifteen-minute presentation to a small board appointed to review the value of the different groups.

How will you explain the value your industry and particular company bring to the community?

Let's say you're a lawyer. As an attorney, what do you do that is so valuable that it would inspire venture capitalists to invest in expanding your firm? Could you say that you defend the innocent, protect the at-risk population and hold paramount the value of human life?

You take complex situations where people are injured in some way, then shape and communicate these complexities in an easy to understand way—while making emotional connections for the people involved.

You are a purveyor of justice. You work hard to protect the rights of the people and defend the philosophy and truth that your forefathers and the founders of this country established over 230 years ago.

Without you, criminals would fill the streets; there would be chaos and disorder everywhere making the world a dark and scary place to live.

In contrast, because of the work you do fighting constantly for justice and the rights of the others, we have a system of checks and balances that allows mankind to not only survive, but indeed thrive in their chosen endeavor.

If I were on this board, and I heard this, I'd cut you check on the spot.

Make this type of speech with sincere passion, enthusiasm, and vision to every employee that works at your company and they will give you far more than the value of the one-time gift this appointed board would have given.

I've worked for some visionary men and women who know what it's like to cast the vision and make it more enjoyable to work on even the simple day–to-day projects. Your team will be much more engaged if you talk about the *why* instead of the what. Why are we doing what we're doing is so much more important than what we do. Over-communicate *purpose.*

There's a story I heard about some construction workers who were working on a building downtown. A reporter came up to the first worker and asked him what he was doing. He responded in a gruff New Yorkers accent, "I'm pouring some concrete – you gotta problem with that"?

A little shaken up, he proceeded to a second worker and asked what he was doing. He responded, "I'm building a wall for a 5,000-square-foot building." A little better, but his answer still didn't tell the reporter much of what he was looking for.

So, he went to a third worker and asked him what he was doing. He responded, "I'm building an orphanage to serve the homeless kids in this neighborhood. There are so many kids out on the streets right now that we want to provide a safe and loving environment for them to grow and call home. We're going to change their lives."

All three were technically correct, but the third had the vision and clearly knew the purpose of what he was doing.

Who would you want to follow?

I was recently making a presentation to our City Council on a project that I'd been working on for nearly five years. We were reaching a major decision point on this important fifty million dollar multifaceted project. A majority of our Council had just been elected and quite frankly didn't know all the details that staff had been working on over the past few years.

As such, the City Manager decided it would be a good idea to have our team make a formal presentation to inform the Council on some of the project details, including the financial situation. Two-and-a-half hours later, concluding our presentation by sharing the reality of an eight million dollar short-fall, the Councilors started suggesting additional funding sources, grants, and financial partners who might be interested in supporting the project.

Why would their perspective (attitude) be to find more money versus simply rejecting the idea and reducing the scope of the project? Because the vision and the transformational qualities of a legacy-type project were bigger than the obstacles that needed to be overcome to make it happen. I guarantee we would postpone a few half-million dollar projects in order to make up the eight million dollar funding gap on this project—simply because the vision and importance is much bigger than accomplishing a dozen smaller projects.

In comparison, looking back at the five points in Kennedy's space speech,

1) we **set huge expectations** to have a signature transit station the Regional Transportation District would leverage for years to come.

2) We **gave credit** to the many funding partners, city leadership and current City Council.

3) We **identified the competition** with other stations that would serve the area.

4) We **appealed to a higher cause** of serving a lower income part of town and helping raise the property value of people who currently live in the area.

And,
5) established the **risk of inaction** by showing them pictures of the plain looking station that would be inaccessible by the lower-income neighborhood and more dangerous than the one we were proposing.

I can't stress the importance of showing someone the vision of what could be and not just focusing on the nuts and bolts of the project itself.

-

"WHERE THERE IS NO VISION, THERE IS NO HOPE."
– GEORGE WASHINGTON CARVER

-

Consider what happened when God talked to Moses about the Promised Land. God told Moses, *"I have prepared a land for you – a land flowing with milk and honey."* Coming from a steady diet of manna for breakfast, lunch and dinner, milk and honey must have sounded like eating on a cruise ship for the rest of your life.

When casting the vision, it's important to focus on the end result, not the work associated with the goal. In Moses' case, what's important isn't what God said, but rather what He didn't say. He didn't tell Moses there was a land full of cows and bees.

A land full of cows and bees just sounds like a bunch of work. Feed the cows, milk the cows, and clean up after the cows. Take the honey from the bees, get stung by the bees, and take more honey from the bees. You get the picture. But a land flowing with milk and honey doesn't sound like work, it sounds like paradise.

The same is with any project – you have to cast the vision.

Going back to the project example, we gave the Councilors a vivid image of what the completed project would look like by telling the story of about what could happen if the funding went through.

We asked them to imagine a warm summer evening as the sun began fading behind the clouds and cool air swept across the lake, making subtle wrinkles in the surface of the water. Birds dip and dive, catching mosquitoes that would otherwise tempt a fish or two to surface out of the safety below to catch a nibble. Around the other shoreline you see some kids holding hands and running in circles, while others play tag while splashing through the creek.

Waves of bicycles seem to flow on the never-ending trail that meanders through the park, while still others stroll quietly on the crusher, fine trails that weave gracefully in and out of the lake's edge. Park benches are full of patrons sipping on a frappuccino or enjoying some ice cream. A random group of violinists decide to serenade passersby with a recognizable piece by Mozart.

As the sun and the music begin to fade, you see the last train of the evening departing the illuminated station. You begin to smile as if knowing that somehow, this day couldn't be more perfect. And you have the feeling that you've just experienced a surreal moment when the heavens came down and kissed the Earth and you had a front-row seat.

Can you feel it? Imagine being one of the Councilors – wouldn't you get excited about such a project?

The biggest mistake most engineers and other technical people make in these types of presentations is they get too far into the details of the project. I've seen groups of people drop into a deep narcoleptic sleep when I started talking about projects in this way. If we're honest, most people really don't care about the details. But they care a whole lot about how it impacts them. If there is one thing you get from this book, let it be this – *sell the sizzle, not the steak.*

Here's what I mean. The other night, my family and I went to Chili's for dinner. We had just sat down and were contemplating what to order when all the sudden, out of the corner of the restaurant, I heard a faint sizzle. It got louder and louder, and as the waiter walked past our table I began to smell that wonderful aroma of grilled steak, onions, peppers—all adorned with a medley of guacamole, cheese, salsa and sour cream. This delectable dinner consumed my brain and for the next few minutes I only had one thought on my mind, "get me some sizzle!"

Without hesitation, it was a plate of sizzling fajitas for me.

The same goes with any presentation that you have to make – sell the vision and excitement of what the end product will do for the client—not all the details and obstacles that you'll need to overcome. The only reason you're getting the interview is because you are an expert in your field. When everyone else being interviewed is an expert as well, you need to find something to set yourself apart. That's when having a focused vision for what will be is the key to success.

> **WHATEVER PROJECT YOU ARE WORKING ON – IF YOU WANT PEOPLE TO BUY INTO WHAT YOU'RE DOING, SELL THE VISION, NOT THE DETAILS.**

chapter 3

CONNECTION SPEED

ONE OF THE MOST important qualities a person can have is their ability to connect quickly with other people. If you can't connect with others, you'll spend a lot of time standing alone.

Think of every contact as an exercise in speed dating. In the first thirty seconds the other person will decide whether you are someone he or she will trust and build a relationship with – or if you're just trying to get something from them, while the defenses go up.

Early on in my career I was at a business networking event. I was so anxious to make some contacts for my company that I must have looked like a puppy in a room full of chew toys. I was even salivating a bit and if I had a tail it definitely would have been wagging at a million miles an hour. I'll never forget when I went to introduce myself to one of the gentlemen there. I was a nervous twenty-something, fresh out of college, trying to make a name for myself. With a rapid heartbeat, huge smile, and perspiration collecting on my forehead, I looked at him and with a nervous and squeaky voice I said, "How's it going?" He looked at me briefly, probably wondering if I was overly medicated, and kept on walking.

I stood there for a moment in disbelief trying to wrap my brain around what just happened. I had a couple choices at that moment. I could be irritated at that guy for just walking by me, or I could take responsibility for my own actions and learn. I did a little of both. I smiled and said to myself, "Ok one down, ninety-nine to go." I knew I'd get better at this whole relationship-building thing and, quite frankly, it would just take time and a lot of practice.

At the same time, I didn't want to blow through all 100 potential contacts and have nothing to show for it. So, I decided it would be a good idea to learn how to meet people and build relationships.

But how?

I had just graduated with a Bachelor of Science in Civil Engineering from Colorado State University, I had spent the last four-and-a-half years of my life entrenched in technical manuals, differential equations, statistical analysis, and enough math courses to nearly have a minor in the subject.

But I didn't have a single class on how to get along with others or initiate conversation at a social event. The best instruction I had on connecting with others was bowling class. Let's just say that if I were looking for a degree in how to build professional relationships, my time in the bowling alley didn't even count as a prerequisite.

I began to look around, trying to find out what it takes to connect with people. My roommate at the time, and best friend in college, happened to be a history major. There's something about the liberal arts curriculum that seems to attract the outgoing, personable types a bit more than engineering school.

I always imagined that the first job many liberal arts students get requires them to wear a headset and make statements like, "Would you like to upsize that?" But at least they actually make eye-contact instead of staring at their shoes like I used to do.

He just seemed to have a knack at getting along with people, and people really liked him. I started taking mental notes of how he approached people. What he said, how he said it. It all just seemed to flow without him thinking about it.

We'd go out to different social events on the weekends and he'd be the life of the party, laughing and connecting with everyone around us. He wasn't exceptionally good looking – so that gave me hope. He wasn't wealthy and he wasn't super smart – again, favor was shining upon me.

The one thing he did have was *confidence.* He was so sure of himself that it made everyone around him comfortable with who they were. There was simply no arrogance. He seemed to say to everyone he encountered, "I am who I am, all my flaws and issues, so you can be who you are and we'll get along just fine."

This is the *first key* to connecting with others.

❯❯ BE COMFORTABLE WITH WHO YOU ARE AS THIS WILL LET OTHERS BE COMFORTABLE WITH WHO THEY ARE.

It's so refreshing to be around people who are truly comfortable with who they are. There's a peace about them and you immediately feel as though you can trust them because they don't hide behind a mask of expectations.

-

"TO BE YOURSELF IN A WORLD THAT IS CONSTANTLY TRYING TO MAKE YOU SOMETHING ELSE IS THE GREATEST ACCOMPLISHMENT."
-RALPH WALDO EMERSON

-

When people sense this confidence and feel comfortable around you, they're more likely to open up and start a conversation.

This is the *second key* to making a connection - figure out what you will say to the other person that will start an interesting conversation. In order to engage others in an interesting conversation, you need to ask Interesting Questions. I like to call this conversational IQ.

❯❯ TO HAVE AN INTERESTING CONVERSATION, ASK INTERESTING QUESTIONS (IQ). THIS IS WHAT WE CALL CONVERSATIONAL IQ.

There are many ways to open up a conversation, ranging from talking about the weather to asking the other person what brought him or her to this particular event. There are entire books written on this subject – find something that works for you in every venue. Whatever you end up asking, the ultimate goal is to discover something that is important to the other person.

If it's a business social-function in your industry, you can ask people who they work for or why they decided to come to this particular event. If it's a retirement party, you can ask people how they know the guest of honor. If it's a seminar on a general topic of interest that crosses many industries, you may simply ask people what line of work they are in.

In any scenario, the first thing I like to do when entering the conversation is to be the first to reach out my hand, and introduce myself with a firm handshake. Being the first to introduce yourself puts you in the driver's seat of the conversation. You become a type of host and a person who is interested in connecting with others. Almost invariably, the other person will be relieved that you were the first to introduce yourself to him or her.

Some folks may lean more toward the introverted side or feel as though they don't know anyone there and have no idea how to reach out to others. The simple fact that you introduce yourself first takes all the pressure off of them and allows them the opportunity to connect to you.

I've done this a number of times. Some of those first acquaintances have turned into lifelong friendships. I've literally had people who I've known for years remind me that I was the first to say hello and make them feel welcome. This feeling of belonging is one of the strongest interpersonal needs we have. This simple act of reaching out to others will make a long-lasting connection that won't soon be forgotten.

After you've made this initial connection, the next few questions you ask can make or break the conversation. You'll have a limited amount of time to listen to how they answer the ice-breaker question to determine where their interest lies. Pay close attention to everything they say and how they say it and try to find something – anything— in common with this person.

From the three scenarios above, there's a good chance that you know people in common or someone in their industry. When you find something in common, briefly mention it without hijacking the conversation and bringing it back to yourself. Quickly turn the questions back to them and their interests. This simple yet sometimes challenging approach keeps the interest and focus on the other person.

There's a saying that goes, *"If you want to be interesting, be interested."*

There's a powerful dynamic that happens when you take this approach, as the other person will find you interesting simply because you took a genuine interest in them.

The same is true with relationships that you've already established and you want to keep building.

The ***third key*** to connecting with others is to speak in terms of the other person's interests.

When I first started working for a local government, the City Engineer would frequently take me with him during marketing calls from engineering consultants looking for future project opportunities. Since he'd been working for the city for nearly thirty years, he had come to know every local engineering company.

He would often reminisce about the founders of certain companies, when they started out, and how many years they stayed. Some of the firms had been acquired by other companies and worked through several name changes—but the people were the same. Whether the logo on their lapel had changed or not, they still knew Dave and that was often the key to their continued success.

Invariably, the consultants who had gotten to know Dave well knew that in his off time he and his wife participated in dog shows in hopes of presenting a national champion. In time, they were invited to the national championships for their breed of dog.

Every time we met with one of these well-known consultants, the very first question they'd ask Dave is, "Have you been to any dog shows lately?" This would lead to at least ten minutes of jovial conversation, as Dave would have some funny story to share about their latest trip or how well one of his dogs placed in the competition.

Now, the question you may be asking is whether or not these consultants were really interested in dog shows or if they were just looking to get a project. The answer is both. If you simply start asking questions about the project to begin the conversation, then you've missed the opportunity to connect.

Truthfully, you need to find out what makes your client *tick* before you start *talking*.

-

"NOBODY CARES HOW MUCH YOU KNOW, UNTIL THEY KNOW HOW MUCH YOU CARE."
- THEODORE ROOSEVELT

-

Feigned, superficial interest is worse than not even asking the question at all. It's like asking someone about their weekend only to interrupt them to begin talking about how amazing your weekend was and totally hijack the conversation. The generous thought is totally lost in your selfish actions.

Over time, I have seen firsthand the proper way to market potential clients and the last thing you want to do is dominate the conversation about how interesting you are.

Remember back in the introduction chapter, the three architectural teams that we interviewed? The winning team was focused on how great the client was, not how great they were.

❯❯ FOCUS ON HOW GREAT YOUR CLIENT IS, NOT HOW GREAT YOU ARE.

Like it or not, most clients know exactly who their go-to consultants are for projects. This isn't something that happens overnight, but is something that develops over time by consistently working hard on building the relationship.

Not just a one-time activity, but on-going interactions coupled with delivering high-value services—time and again.

The *fourth key* to connecting is simply to be approachable.

There is an epidemic in our society right now. People are so distracted with the constant flow of information that they've completely forgotten how to connect with each other.

Unplug and connect!

I was recently in a meeting with only five other people, so it was easy to observe the situation. Four of us were engaged in a topic of interest while the fifth was evidently checking social media when all of the sudden this person abruptly interjected a "recent event" that had nothing to do with the topic at hand. The sensationalized story was fully intended to draw out emotion in its readers, which it did, and worse yet, the story was a lie. So now, the entire conversation was taken off course while we had to deal with this interruption.

This case is obvious, but hundreds of other situations go more unnoticed as distracted people keep their thoughts to themselves. But do they really? Our minds were created with synapses that constantly connect information to our emotions. When you read about a sad story, you might as well be experiencing it live because you will likely have the same general, emotional response.

When you're in a jovial conversation with friends and read a sad post on your smart phone, your mind doesn't differentiate between what you read and what you said — it just responds by connecting the story to the emotion. This results in an awkward and sometimes inappropriate emotional response that completely separates you from connecting with the other person.

You also need to be cautioned on how susceptible you are to the information you receive. In 2012, Facebook, in collaboration with researches from Cornell and the University of California, San Francisco, conducted an experiment with almost 690,000 of its users. In this experiment, Facebook tweaked the news feed showing some people a higher number of positive posts while others received a higher number of negative posts.[1]

Not surprisingly, the users who received more negative content were more likely to post negatively while the positive group reciprocated with uplifting comments.

Knowing how susceptible we are to bad information, it's more important now than ever to fact-check everything you read and hear.

If you still have an issue and potential addiction with constant news feeds on your smart phone – we have an app for that. There's an application that you can buy call "Anti-Social" that is designed specifically for you. It strategically locks you out of your social networking sites for a predetermined amount of time that you set every day. This way, even if you wanted to check your status updates and feeds, you won't be able to until the designated time.

You might consider this if you can't put your phone down every now and then. For some of us, it's for our own good.

◆❯ IF BUILDING THE RELATIONSHIP IS IMPORTANT TO YOU, THEN PUT DOWN THE PHONE – UNPLUG AND CONNECT.

The *fifth key* to connection is to realize that just because you're talking it doesn't mean that others are listening.

We judge ourselves on our good intentions and we judge others by their actions. This attitude couldn't be more true for me, and it came back to bite me on a day that I'll never forget.

As an engineer working for a local government, I find that sometimes I spend as much time on the "other duties as assigned" category of my job description as I do on my primary functions. One such duty is making educational presentations to local-area school-children. Thankfully for me, there is an amazing group of professionals who have come together over the years and assembled a full day of educational activities for around 1,500 fourth and fifth graders from area schools. Three different cities combine efforts to make this whole project come through, and the kids are at a great age to connect with the speakers.

I happened to be scheduled to teach five different classes of fifth graders throughout the day on storm-water quality. This is such a fun age to teach as the kids are totally engaged as they still thrive on learning as much as they can.

That morning, I had just finished getting the full-scale model ready to go for the presentation. I was spending a bit of time in the hallway reading a variety of the American history boards that blanketed the walls. There were photos of various historical events from the Great Depression, Dust Bowl, and World War II. As I was reading one of the captions I noticed that a young lady was walking toward me down the narrow hall. Thinking that she was also one of the presenters for this event I continued looking at the photos while making the comment "These pictures are amazing, I could spend all day looking at them."

Without turning to make eye contact, I expected that out of courtesy of my engaging commentary that she would respond with something similar like, "Me too, I'm amazed at the quality of these pictures and the stories they tell." But alas, she made absolutely no comment at all and just kept walking by.

Immediately I thought to myself that she was a self-indulged individual or more likely just another introverted engineer that trembled at the fear of actually talking directly to someone. Thankfully, I didn't get too depressed over her lack of interaction. I just chalked it up to her casual indifference and proceeded to my assigned classroom to begin my presentation.

My great epiphany came about an hour later when I happened to be walking down this same hall. She was walking directly toward me but this time another lady was with her. What happened next opened my eyes to see and feel, in an instant, a sense of remorse, compassion, guilt, and yet another opportunity to enjoy the bitter taste of some humble pie. You see, she was speaking in sign language with the lady next to her. Just like that, I could now understand what I could not see. She wasn't being rude to me – she simply didn't hear me – because she was hearing impaired.

How often do we think we are communicating with others and let ourselves get all bent out of shape because the other person didn't respond the way we thought they should respond? I can think of all the ways that I could have treated this particular lady if I didn't happen to see her signing to her friend. I probably would have ignored her and maybe reciprocated with a little arrogance or rudeness of my own, making a negative judgment upon her. And she would have been completely oblivious to the reason why I was acting this way.

My good friend Joe shared a similar situation that happened to him over the course of a few months.

I'll let Joe tell the story himself:

"I try to get myself into the gym in the mornings before work and tend to see the same faces day – in and day – out. This is generally a time that I spend alone in my thoughts, but there is usually a little chitchat that goes on with the other regulars.

There was this one guy who rarely smiled and never engaged in any of my halfhearted pleasantries. His rejection of my presence all came to a head one day when we were both sitting in the sauna together and I asked how he was doing. In response to my inquiry, he just got up and walked out without saying a word. I was thinking, what is the matter with this guy? He really needs to lighten up!

A week or so later, I happened to be entering the gym just before this same guy. I held the door for him and he said "thank you". However, as he uttered these words, I noticed that his words were broken as a result of what seemed to be a significant speech impediment. So his reluctance to engage in conversation was not because he had an attitude problem. He just didn't want to have to deal with his impediment during his personal time at the gym. At that realization, it was painfully obvious that all my previous negative judgments towards this guy were based on my own misconceptions, which turned out to be completely unfounded. I try to be slow to judge people, but in this instance I definitely failed.

I believe that often times we are quick to judge, which can be perceived as an arrogant character trait. For one, sometimes we feel that we can figure out a person based on a snapshot in time. The fact that we are judging at all suggests that we don't have any flaws of our own. This is both arrogant and wrong. At our root form we are all created equal, so shouldn't our first reaction be to embrace, rather than cast-off with quick judgments?

Although we all have character flaws, I feel strongly that the human spirit is good. That is why I think it is so important to let our own positive outlook shine through and to do the best we can to extract it out of those around us. This not only puts us in a good place, but also puts those around in a good place. The great thing is that I don't think that this philosophy has any boundaries. Be it family, friends, coworkers, clients, or whomever, I have seen it work on many levels. This is not a magical power, we all have it in us and we simply need to choose to activate it. Positivity breeds positivity no matter who you are or who you are with."

In this day and age, with multiple venues for communicating with others, it's easy to think that just because you are talking, others are listening. There have been emails and voice messages that I've sent without any reply. On a few occasions, it's turned out that the other person never got my message due to some technological issue.

I try really hard to be slow to judge others.

-

"THE SINGLE BIGGEST PROBLEM IN COMMUNICATION IS THE ILLUSION THAT IT HAS TAKEN PLACE."
– GEORGE BERNARD SHAW

-

❯ BE SLOW TO JUDGE AND QUICK TO ASSUME THE BEST.

The *sixth key* to connecting with others is to apologize emphatically when you've made a mistake.

We were interviewing a design-build team that was assembled to build an eight million dollar parking structure. The process was complicated as we were trying to mix a parking garage contractor with a mixed-use developer that would build retail and residential uses right next to the garage. These are different product types and teams that came together to propose on this exciting project.

During the selection process, we had an unsolicited bid come in that wasn't part of the final two groups we were interviewing. But given their initial interest in the project and the timeliness of the inquiry, we allowed them the opportunity to make one last pitch to our committee.

These individuals and their companies were and continue to be well-respected in the construction industry. The reason I preface this story with that statement is even though they were stars as individuals and companies, they were a disaster as a team.

The project lead introduced his team and the various disciplines and areas of expertise they all brought to the table. Every one of them had impressive resumes and recent successful projects of similar nature.

The lead project manager began by sharing a concept plan that was similar to the vision the city staff had for the project. But then, about half way through their presentation, one of the other architects began to share his vision on how to lay out the site. Sitting on the client side of the table it became evident quickly that he hadn't really cleared this vision with the rest of his "team." He was clearly out on a creative island all by himself without a dingy to bring him back to the rest of his team. Worse yet, his vision was in complete opposition to the client's vision for the site plan.

That said, after patiently and graciously listening to him for a number of minutes, the Department Head finally cut him off and simply said "this doesn't work for us."

At this point, the architect had a choice to make. Take the suggestion and offer to relinquish his idea and embrace the city's vision, or dig his heels into his ideas and alienate himself from the rest of the team.

You can guess which one he chose. That's right, he dug his heels in and quickly exited the building right as the meeting was brought to a close, without shaking hands and thanking the team for their time.

I was so shocked at this professional's actions that I began to perform a simple relational postmortem on the situation. Though I don't know all the intricacies of what he may have been going through, I did ask myself what he could have done to save the relationship.

If he would have read the faces of our team and accepted the direct criticism that his plan wouldn't work for us, he could have saved face by honestly and emphatically apologizing for having such a contrarian approach to the project.

All he needed to do was say something like, "Oh my, I must be totally off-base here. Sorry about that! After listening to you I now realize were are not on the same page, so I better go back to the drawing board."

❯❯ WHEN YOU MAKE A MISTAKE – ACKNOWLEDGE IT.

Had he done this, I can almost guarantee the committee would have accepted the apology and felt empathetic toward his situation. They may have even encouraged him to come up with some new ideas. Instead, he put his ideas before the client's interests and lost the project.

It doesn't matter how great your ideas are – if you don't have a buyer, they don't have a value.

Remember, at the end of the day it's the relationships that matter most to us. All projects come to an end. *But the relationship you build through the project will lead to more and more projects.*

Never lose sight of the most valuable aspect of any project – the relationships you cultivate and nurture.

chapter 4

POSITIVE ATTITUDE

-

"A GREAT ATTITUDE DOES MUCH MORE THAN TURN ON THE LIGHTS IN OUR WORLDS; IT SEEMS TO MAGICALLY CONNECT US TO ALL SORTS OF SERENDIPITOUS OPPORTUNITIES THAT WERE SOMEHOW ABSENT BEFORE THE CHANGE."

- EARL NIGHTINGALE

-

The ability to connect quickly will get you in the door. The power of using a positive attitude will keep the door open.

There is something so attractive about people with a positive attitude that you are magnetically drawn to them no matter what the situation is. Just being in their presence gives you the same feeling as walking into a home where they've just baked a batch of your favorite cookies. Your host doesn't have to say anything, yet you feel warm and appreciated simply by being with them.

The great thing about a positive attitude is that it's a choice. Every time you meet with your client or the project team, you have a choice to make. Do you become the life giving sunshine in the meeting, or do you carry around a bag of smelly garbage from the day's issues and offer to share it with the rest of the group?

How many meetings do you sit through where the other participants can't wait to share all their garbage with you? What do we often do in return? Pull out some of our own garbage and offer to share a little with them. Soon, everyone is trading their refuse.

There's a story about a taxi driver who was driving a man from the airport to his hotel room in New York City. As they were driving along a car next to them cut right in front of the taxi cab nearly running him off the road. The other driver started yelling and waving his hands at the taxi cab driver as if somehow it was his fault he nearly got hit. The passenger in the taxi cab was waiting for the cab driver to start yelling and giving a few select hand signals of his own, but instead he simply slowed down, smiled, and waved at the reckless driver in front of him. The passenger was in complete dismay as he witnessed this whole thing take place. He asked the taxi driver why he didn't start screaming back at the other driver. His profound words made a huge impact on the passenger. He said, "I just visualize that everyone carries a bunch of garbage with them on the road looking for a place to dump it. But as long as I stay calm and collected, not blowing my lid off, they can't leave it with me."

So true. Said another way, if you're tired of people getting your goat, keep it tied up.

This is a critical attitude to have on any project team when things go wrong. The way you react to those problems will leave an indelible impression on the rest of the team.

Working in the civil engineering field for over twenty years now, I've never designed or managed a project that from start to finish didn't change in some way. Some changes were relatively insignificant – others were huge. How many times have you had a client say something like, "Can you just move the building a few feet east, and mirror it so the office is on the north side instead of the south?"

Seems like a reasonable request right? Well, not for the architect, engineer, plumber, electrician, drafter, or anyone else who has already started and possibly finished their design work. There's usually a lot of posturing during a project when no one wants to be the first to start the final design because change is inevitable. The later the change occurs, the more it costs everyone involved in both time and money.

At the same time, change is an absolute must on every project to get the best possible design that will last long after the time it took to make the change. It's how everyone responds to that change that makes the difference.

Currently, I'm working on the most exciting projects of my career – The Westminster Station. This is a multi-faceted project that includes a commuter rail station, a parking structure, new roads, bridges, regional bicycle trail, utility relocations, property acquisitions, regional flood detention facility, and a redevelopment plan that covers more than 160 acres of land. The entire infrastructure associated with this project totals well over 180 million dollars.

By far, this is the best time I've ever had working as a civil engineer. Why is that? Because it brings with it huge challenges and obstacles to overcome from budgets, politics, construction coordination, and managing constant change. Part of what makes it a total blast is the talented and competent professionals that take an active role in problem-solving and creating solutions. It's their positive attitude toward these changes that makes the difference.

We were recently struggling through one of the architectural elements for the station and had come up against a significant challenge on how to make the feature work. The conceptual renderings were great, but when we started to integrate them into actual space there were some issues ranging from structural, safety, cost, long term operations and maintenance points of view. After giving the issues some detailed thought, I brought an idea to the team that would solve most of these issues. The only problem was that it completely changed the signature architectural element.

As I presented the idea to the design-build team, I was met with a bit of contemplative silence. The solution was sound, but it brought with it a few other nuances—the biggest of which was the divergence from the architectural style.

As I mentioned, this team is made up of rock stars, so, instead of throwing out the idea, they quickly started offering up ideas on how to make it work. The team spent the following week working through the design and coming up with new renderings and cost estimates. In turn, I took this information with the detailed cost estimates and solicited input from a couple key folks from the city. We agreed it was the right change and authorized the designers to advance the concept.

The challenge came a couple weeks later when I presented this change to the larger stakeholder group at the city. With one quick glance our city attorney simply said, "This really changes the look that's been shared with city council and the public over the past two years." His point was well taken. Though to be honest, it put me in a really tough position. Now I had to go back to the design team and say, "Sorry folks, but we need to go back to the other design."

This is where the positive attitudes of the design team really came through.

They could have been resentful that two weeks went by and designs, cost estimates, construction schedules, and materials procurement were all in progress. However, as I addressed the team – with a huge amount of empathy for the impact this would have on them – they were quick to offer up new solutions to make the signature element work.

It's this can-do attitude that draws out the best in the entire team. Because they kept an open-minded attitude, we were able to discuss more details and opportunities that now existed that we didn't see before. Having a positive attitude is a "free" way of adding value to every team you with whom you work.

I've been part of other design efforts that when this sort of change occurs there is a quick round table discussion to identify the chain of blame. Every single time this has happened it's been a complete waste of time and energy and doesn't bring any value to the project.

❯❯ HAVING A POSITIVE ATTITUDE ENHANCES CREATIVITY AND ADDS SIGNIFICANT VALUE TO EVERY PROJECT YOU WORK ON.

In addition to bringing more creativity, a positive attitude reduces your stress.

The number one killer of Americans is heart disease and one of the primary contributors to that disease is stress.

There are various studies that link a positive attitude to living a longer, healthier life. A recent article published by the Mayo Clinic made the following observations:
Health benefits that positive thinking may provide include:[1]

- ☒ Increased life span
- ☒ Lower rates of depression
- ☒ Lower levels of distress
- ☒ Greater resistance to the common cold
- ☒ Better psychological and physical well-being

☒ Reduced risk of death from cardiovascular disease

☒ Better coping skills during hardships and times of stress

We all know how important it is to have a positive attitude, yet from time to time we all find ourselves getting caught in a spiral of negative self-talk. This is normal as our minds will follow the path of least resistance when left unattended.

Liken your mind to a garden. You can plant the seeds of greatness but if you don't pull the weeds of negativity, after a while you won't see anything but the self doubt, guilt, lost opportunities, etc.

In "The Strangest Secret,"[2] a famous essay written by Earl Nightingale, he shares an analogy of the farmer and the field:

AS YE SOW — SO SHALL YE REAP.

The human mind is much like a farmer's land. The land gives the farmer a choice. He may plant in that land whatever he chooses. The land doesn't care what is planted. It's up to the farmer to make the decision. The mind, like the land, will return what you plant, but it doesn't care what you plant. If the farmer plants two seeds — one a seed of corn, the other nightshade, (a deadly poison) then waters and cultivates the land, what will happen?

Remember, the land doesn't care. It will return poison in abundance as it will corn. So up come the two plants — one corn, one poison.

As it's written in the Bible, "Whatever you plant is what you're going to harvest."

The human mind is far more fertile and mysterious than any piece of land, but it works the same way. It doesn't care what we plant, success or failure, concrete, worthwhile goals or confusion, misunderstanding, fear, anxiety, and so on. But what we plant in it will return to us.

The problem is that our mind comes as standard equipment at birth. It's free. Things that are given to us for nothing, we place little value on. Things that we pay money for, we value.

The paradox is that exactly the reverse is true. Everything that's really worthwhile in life came to us free — our minds, our souls, our bodies, our hopes, our dreams, our ambitions, our intelligence, our love of family and children and friends and country. All these priceless possessions are free.

But the things that cost us money can be replaced at any time. A good man can be completely wiped out and make another fortune. He can do that several times. Even if our home burns down, we can rebuild it. But the things we got for nothing, we can never replace.

Our mind can do any kind of job we assign to it, but, generally speaking, we use it for little jobs instead of big ones. So decide now. What is it you want? Plant your best goals in your mind. It's the most important move you'll ever make.

Do you want to excel at your particular job? Do you want to go places in your company or in your community? Do you want to obtain what you need? All you have to do is plant that seed in your mind, care for it, work steadily toward your goal, and it will become a reality.

There's a law like the law of gravity. If you get on top of a building and jump off, you'll always go down — you'll never go up.

It's the same with all the other laws of nature. They always work. They're inflexible. Think about your goal in a relaxed, positive way. Picture yourself in your mind's eye as having already achieved this goal. See yourself doing the things you will be doing when you have reached your goal.

Every one of us is the sum total of our own thoughts. We are where we are because that's exactly where we really want or feel we deserve to be — whether we'll admit that or not. Each of us must live off the fruit of our thoughts in the future, because what you think today and tomorrow, next month and next year, will mold your life and determine your future. You're guided by your mind.

We become what we think about. A person who is thinking about a concrete and worthwhile goal is going to reach it, because that's what he's thinking about. Conversely, the person who has no goal, who doesn't know where he's going, and whose thoughts must therefore be thoughts of confusion, anxiety, fear, and worry, will thereby create a life of frustration, fear, anxiety and worry. And if he thinks about nothing ... he becomes nothing.

-

"A MAN'S LIFE IS WHAT HIS THOUGHTS MAKE OF IT."
– MARCUS AURELIUS, ROMAN EMPEROR

-

I had a mentor of mine share with me the process of protecting my mind from the negative influences of my surroundings. This simple process follows the pattern of "Input – thoughts – action – habit – character – destiny"

Knowing this pattern is the key to all success. Information is transferred through our minds through chemical or electrical synapses. Pathways are created over time that speed up the transfer of information through our minds which results in faster reaction times.

In the same way the best athletes in the world practice the same motion over and over again so they gain muscle memory, our minds work hard every day creating a reaction to the input it receives. Over time, our minds create habits that are harder and harder to change. One of the best ways to curb the habit is by changing or stopping the input in the first place.

Here are a few ways that you can change the input and help develop the positive attitude that will make you a valuable player on any project team that you are part of and, in turn, help you win more and more projects!

1. Create a list of positive or inspirational quotes that you read every day.

2. Surround yourself with pictures of family and friends sharing good memories.
3. Subscribe to substantive blogs/newscasts/articles – Don't take the path of least resistance and fill your day with the local/national news media. You're guaranteed to fill your mind with negativity.
4. Listen to motivational and inspiring stories.
5. Share something good with someone else every day.

-

**"YOUR BELIEFS BECOME YOUR THOUGHTS,
YOUR THOUGHTS BECOME YOUR WORDS,
YOUR WORDS BECOME YOUR ACTIONS,
YOUR ACTIONS BECOME YOUR HABITS,
YOUR HABITS BECOME YOUR VALUES,
YOUR VALUES BECOME YOUR DESTINY."**

– MAHATMA GANDHI

-

Doing these things on a daily basis will help you become more creative and solution-oriented on every project that you work on. Every obstacle has an equal or greater solution that is waiting to be found. You're ability to deploy a positive attitude is the key to how you get to that solution.

Joe Pleban from Virginia was facing the impending amputation of his lower left leg.[3] The twenty-three- year-old had a rare ankle disease that was causing so much pain that he wasn't able to live life doing the things he enjoyed the most without excruciating discomfort. As such, he decided amputation was the only way he was going to get his life back. Instead of filling his mind with the sorrow and unfortunate hand he had been dealt, Joe decided to have some fun documenting his final adventures with his leg. He took his leg to the Caribbean, to play paint-ball, sky-diving, got a tattoo on his lower leg with a dashed line that read "cut here." His attitude became a world renowned story. Why? Because he had such a great attitude about the pending amputation.

-

"THE LONGER I LIVE, THE MORE I REALIZE THE IMPACT OF ATTITUDE ON LIFE. IT IS MORE IMPORTANT THAN EDUCATION, THAN MONEY, THAN CIRCUMSTANCES, THAN FAILURES, THAN SUCCESSES, THAN WHATEVER ANYONE MIGHT SAY OR DO. IT IS MORE IMPORTANT THAN APPEARANCES, GIFTEDNESS OR SKILL. THE REMARKABLE THING IS THAT WE HAVE THE CHOICE TO CREATE THE ATTITUDE WE HAVE FOR THAT DAY. WE CANNOT CHANGE OUR PAST. WE CANNOT CHANGE THE WAY PEOPLE ACT. WE CANNOT CHANGE THE INEVITABLE.

THE ONE THING WE CAN CHANGE IS THE ONLY THING WE HAVE CONTROL OVER, AND THAT IS OUR ATTITUDE.

I AM CONVINCED THAT LIFE IS TEN PERCENT WHAT ACTUALLY HAPPENS TO US AND NINETY PERCENT HOW WE REACT TO IT."

- CHARLES SWINDOLL

-

Start today. You have nothing to lose and everything to gain.

chapter 5

UNEXPECTED GENEROSITY

NO MATTER WHAT endeavor we're pursuing to make a living, it doesn't stop at five p.m. In fact, I'm sure that when I shut off my devices for the night, the email messages/blog posts/twitter feeds/Facebook status updates will somehow multiply like a bunch of rabbits while I sleep. When I wake up in the morning, I will spend the first ten minutes of the day skimming the subject lines and deleting or ignoring eighty percent of them.

I'm too busy to spend time on the things that I want to; because I spend most of my time on the things I have to do. I'm always looking for ways to increase the amount of time I've been given on things that I want to be doing.

Here's one solution: *Give your time away.*

That's right. *Give your time away by investing it in others.*
There was a study done by the University of Pennsylvania, Yale University and Harvard University where they ran a series of experiments trying to find a solution to the feeling of not having enough time.[1] Though no one is really given more time in a day than anyone else, there is a feeling people have as to the amount of time they really have to get things done.

In this series of experiments, the scientists found an interesting conclusion. Those who gave more time away actually felt as though they had more time.

"We compared spending time on other people with wasting time, spending time on oneself, and even gaining a windfall of "free" time, and we found that spending time on others increases one's feeling of time affluence. The impact of giving time on feelings of time affluence is driven by a boosted sense of self-efficacy. Consequently, giving time makes people more willing to commit to future engagements despite their busy schedules."

The same is true about expending energy or exercising. When you work out and expend energy, you actually get more energy in return.

In a 2006 article on WebMD[2], research suggested that regular exercise can increase energy levels even among people suffering from chronic medical conditions associated with fatigue.

"A lot of times when people are fatigued, the last thing they want to do is exercise," says researcher Patrick O'Connor, PhD, in a news release, "But if you're physically inactive and fatigued, being just a bit more active will help," says O'Connor, co-director of the University of Georgia exercise psychology laboratory, in Athens, Ga.

In this study, published in Psychological Bulletin, the researchers analyzed seventy studies on exercise and fatigue involving more than 6,800 people.

"More than ninety percent of the studies showed the same thing: Sedentary people who completed a regular exercise program reported improved fatigue compared to groups that did not exercise," says O'Connor, "It's a very consistent effect."

The average effect was greater than the improvement from using stimulant medications, including ones used for attention deficit hyperactivity disorder (ADHD) and narcolepsy.

I was sharing this information at a training I was doing for the American Public Works Association Emerging Leaders course in Denver when one of the attendees came up to me after class and shared his experience. He said that over the past couple years he was busier than ever, but he enjoyed doing yoga at a local gym. Within the first two weeks of joining the class, the instructor quit and so he began volunteering to lead. At first he was reluctant to continue leading the class because of his current workload, but, over the course of a few months, he was full of energy, and, because he was looking forward to leading the class, he became more effective in other areas of his life as well.

He couldn't quite put his finger on it, but after I shared the results of this research, it dawned on him that because he was giving his time and energy away, he really did feel like had more time and energy to take on anything.

❯❯ IF YOU FEEL TIME STRAPPED AND OUT OF ENERGY, GIVE SOME TIME AND ENERGY AWAY!

Another area where I believe generosity will give you more in return than you invest is in the area of knowledge.

We were interviewing the top three consultants who were short-listed to interview on a large drainage project. All three consultant teams had prior experience with the city and two of them had recently worked specifically on earlier phases of the project. To be honest, I came into the interviews thinking that one of the two firms with recent project experience would be awarded the project.

But that didn't turn out to be the case.

The first team happened to be led by a project manager we had worked with before, but their company had been recently acquired by a larger firm in the area. That wasn't an issue as it's typically more important to me to work with the project manager that did a good job than the company they represent. I could tell there were some egos at play during the interview as the gentleman I had built a relationship with sat quietly during most of the presentation while the new company's project lead did all the talking. What a mistake. Even though the new guy was qualified and knowledgeable, he completely neglected one of his strongest assets – the man who had spent the last decade building a relationship with me.

Needless to say, at the conclusion of their presentation I wasn't impressed.

❯ ESTABLISHED RELATIONSHIPS ARE THE STRONGEST ASSETS OF ANY TEAM.

The next group did a much better job. Their presentation was thorough and the team was unified in their approach to the project. They had just finished a smaller project for me in the same area, had all the base mapping, knew the issues we were trying to solve, and could quickly move into the preliminary design phase.

Everything was going well until the very end of the interview when they said one of the most destructive lines I've ever heard, "We know how to solve this problem for you, but you'll need to hire us to find out the answer."

I'll never forget the feeling that gave me. At first I respected their position. This was intellectual property that we would have to pay for. At the same time, the project wasn't rocket science and the solution wasn't that complicated. So they didn't really have much to gain by making that statement.

⊗ BE GENEROUS WITH INFORMATION AS IT HELPS BUILD TRUST.

Pause there for a moment, let's fast forward to the last interview.

The final consultant team was brand new to me. I hadn't worked on any project with them and had only heard a limited number of things about them. Strike one. They could have done a much better job of getting to know me prior to the meeting, so I didn't have high expectations for the presentation or their ability to win the project. I was feeling quite distressed thinking about which of the first two I would hire for the work when neither of them did a very good job in the interview.

What happened over the course of the following hour was interesting to say the least. The lead presenter came across as knowledgeable about the project area and had spent a considerable amount of time on planning. He also did a fair amount of "free" research just to prepare for the interview. He wasn't overly personable and didn't come across as a strong salesman by any means.

What he did do, however, was *completely focus on our project needs.* If you recall from the introduction, winning teams focus on their clients, not themselves. The same was true with this interview. This consulting firm was all about the project. I never heard things like, "We are a great consulting firm and you'd be happy to work with us". Instead what I heard was, "this is an exciting project and we couldn't be more enthusiastic to be here and to be considered for this job."

The final thing that happened made selecting this firm a slam-dunk. This team also concluded their presentation by saying, "We know how to solve this problem…" In this case, however, they finished that sentence with, "…and here's the answer."

This unexpected generosity immediately won over the interview team. We took only a few minutes to review our independent notes and concur that this final team was the one we wanted to work with.

Fast forward a few years later. This design team has secured over a million dollars in fees for this particular project. I'm guessing that was a pretty good return on their initial investment and risk in giving away "free" consulting services.

Many times, however, being generous has nothing to do with the dollar figure associated with the gift you give.

My wife recently had an extended stay in the hospital dealing with an attack from a chronic health issue. After spending the first full day in the emergency room waiting for the doctors to determine a course of action, she was moved to the Progressive Care Unit (PCU) which is just one step better than the Intensive Care Unit (ICU). That night, exhausted and in pain, she wasn't allowed to eat or drink anything until they determined whether surgery was going to be necessary.

Fear and anxiety fill these rooms most of the time. But that night, something unexpected and wonderful happened as I sat by her side. We had a peace about the situation. This wasn't the first time she was in the hospital under these circumstances, though this time the care taken by the doctors, surgeons, and nurses was exemplary. In fact, at some point during the night unbeknownst to either of us, the nurse took the time to fill a Styrofoam cup with some ice chips and a plastic spoon.

Seemingly insignificant, yes, but this simple act of generosity filled the room with love and compassion. At this moment in time, there aren't many options for gift giving. All my wife really wanted was relief from the pain and hope that she would start healing soon.

Because surgery wasn't yet out of the question, she wasn't allowed to have any food or drink. Ice chips, albeit a prelude to water, was the only indulgence she could enjoy. This nurse not only was doing the best job at nursing my wife back to health, but she showed compassion and provided a small, humble treat – yes, ice chips in a cup with a plastic spoon.

Sometimes the most precious gifts aren't the most expensive and unique, but rather they come from the heart putting the other person's interests at the center of the gift.

Staying focused on what the other person wants in lieu of what you want to give is one of the best ways to win with people. These gifts are rarely expensive. In fact, many of the gifts that you can give others are absolutely free, but they cost you some time. For instance, if you want to show a client how important they are to you; focus on giving them some "free" gifts such as:

1. Timely phone calls just to touch base.
2. Return emails and voicemails promptly, even if it's to say, "you're important and even though I don't have the answer right now, stay tuned as I'm working on it for you."
3. Ask them interesting questions and let them talk about the things that interest them most.
4. Listen. Actively listen and ask follow-up questions to learn more without interjecting your own experiences.
5. Turn the project in early and under budget.

6. Provide a little "value added" into the deliverable.
7. Lastly, ask for feedback regularly. People love to help improve others, so let them. Ask, "How am I doing? Please let me know if there's anything else I can be doing for you".

These "free" gifts will help you stay focused on what your client really wants.

Being generous with others also gives you the feeling that others will be generous with you.

Here are two tales that capture the difference between greed and generosity:

A young boy and a girl were playing together. The boy had a collection of marbles. The girl had some candy with her. The boy told the girl that he will give her all his marbles in exchange for her candy. The girl agreed. Secretly though, the boy kept the biggest and the most beautiful marble aside and gave the rest to the girl. The girl gave him all her candy as she had promised. That night, the girl slept peacefully. But the boy couldn't sleep as he kept wondering if the girl had hidden some candy from him the way he had hidden his best marble.

◆ IF YOU DON'T GIVE ONE HUNDRED PERCENT TO THE RELATIONSHIP, YOU'LL ALWAYS DOUBT WHETHER THE OTHER PERSON HAS GIVEN YOU THEIRS.

There once was a young prince who was looking for a lovely maiden to marry and share the future throne with him. He was so well known though his own kingdom that every lady he met would treat him with such reverence that he would never know if she truly loved him. So he decided to journey to a neighboring kingdom where he wasn't known at all.

After spending a few months getting to know the community he found a young lady who truly liked him for his personality and not his portfolio. He felt an undeniable feeling that she was the one whom he would marry and become his princess.

So as to not ruin the surprise and make sure her love was true, he only claimed to be a businessman who was selling his wares in her native country. As they made plans to marry, his excitement grew more and more each day in anticipation of the surprise he had that she would become a princess. She too became more and more excited by his obvious enthusiasm, though she didn't know the reason.
People will respond with the same emotion that is projected to them. If you're excitement is increasing, so will theirs.

In contrast to the first young man, the latter's enthusiasm for the relationship was on the verge of getting better and better.

The first young man would live a life wondering what was being held back.

When you are generous and give more into the relationship than you expect in return, you will find that you always receive more than you expected.

chapter 6

FEARLESS INTEGRITY

-

**"I LOOK FOR THREE THINGS IN HIRING PEOPLE.
THE FIRST IS PERSONAL INTEGRITY, THE SECOND IS
INTELLIGENCE, AND THE THIRD IS A HIGH ENERGY LEVEL.
BUT, IF YOU DON'T HAVE THE FIRST,
THE OTHER TWO WILL KILL YOU"**

– WARREN BUFFET

-

In 2010, like many privately-held companies, the City went through a significant downsizing initiative. This was a painful process but also one that was necessary. I'm pretty sure it was the first time in the city's 100-year history that they had to even considered downsizing.

Through this process, the city leadership initiated something called a "core services evaluation." Every department was required to review their current staffing, workload, and activities they routinely engaged in. Every activity was basically put into separate categories ranging from essential (must have) to quality of life (nice to have).

This was a strategic process to "right-size" the organization into focusing on only what was absolutely necessary before adding on quality of life services. As you can imagine, there were some services the city offered that grew organically over its 100-year history and didn't really fit in with the mission and vision of the organization.

Even though many positions were lost in this process and it's heartbreaking to see good people lose their job, the City didn't have to go back and make further cuts in subsequent years.

In addition to eliminating positions, the City initiated voluntary furlough days where employees could opt to take unpaid time off in order to help the overall City budget. The first person to take advantage of this option was our City Manager with many others following suit including myself. It's much easier to follow a leader who is walking with you versus pushing you from behind.

▶ IF YOU WANT OTHERS TO FOLLOW, THEN LEAD. SET THE EXAMPLE AND STANDARD AND WATCH YOUR TEAM FOLLOW SUIT.

Having fearless integrity running an organization means leading by example and saying no to a lot of good ideas if they don't fit within the core values of the corporation.

Our City management did exactly that – made tough decisions for the betterment of the entire organization.

From this core services process and the annual strategic planning process, the city now has better guidelines from which to make decisions on what programs should be continued and which ones need to be eliminated.

Many corporate decisions can be fairly cold as there are many people on committees helping to make decisions. Usually the ones making the decision aren't the ones who end up losing their job.

It's an entirely different story when the person most impacted is you.

That's where having fearless integrity takes on a whole new life – when you're the one who has everything to lose.

One great example of this was exhibited by Javier Martinez a character in the movie *Courageous*, released in 2011.[1]

Javier was working hard trying to support his family and was being considered for a promotion in the factory that he worked in. During the interview, the owner of the company offered him the job on one condition – that he falsified inventory documents.

He could tell that Javier was struggling a bit with this request, so he asked to think it over and get back to him the next day.

When Javier got home that night and explained the situation to his wife. They were both afraid that he would once again lose his job and have a hard time supporting his family. At the same time, they both knew he had to do the right thing.

The next morning, he went into the owner's office and explained that he couldn't take the job on those conditions. Expecting the worst, Javier was waiting for the owner to kick him out of his office, but instead he reached out his hand and said, "You're hired." He went on to explain that he did this as a test and had already interviewed seven other guys for this position – and they had all failed the test.

This is by far one of the most attractive personality traits of a leader – someone that you can trust completely because their fearless integrity shows through everything else they do.

-

"THE TRUTH OF THE MATTER IS THAT YOU ALWAYS KNOW THE RIGHT THING TO DO. THE HARD PART IS DOING IT."
· - NORMAN SCHWARZKOPF

-

Another example of someone who showed fearless integrity occurred in Hallandale Beach, Florida in 2012 when a lifeguard was fired for saving someone's life.[2] Tomas Lopez was working his normal shift on the beach when he noticed a man that was just outside of his patrol area was struggling in the waters. Tomas took action and saved the man's life. You would normally think he would have received a hero's welcome, but what he got was a pink slip from his company. You see, the man he saved was outside of the coverage zone that he was hired to "guard."

The company that hires lifeguards for this part of the beach has strict instructions to only patrol the specific area of the beach for which they were hired. Though this absolutely makes sense for a number of reasons, the reality is that when Tomas was faced with the choice of saving a man's life or following his company's policies, he made the right choice in saving the man. The choice resulted in the quick firing of Tomas for disobeying this rule.

There are times in life when you need to make the right choice regardless of the personal consequences. When you do, time and your conscience will confirm that you made right decision.

In this case, the company who abruptly fired him came out and made a public apology and offered Tomas his job back. Unfortunately, the damage was done and Tomas would not accept the position.

◆❯ Make the right choice regardless of the consequences.

On the other side of fearless integrity, there are plenty of stories out there about people, who are astonished that their social media profiles could be used as a basis for hiring or firing them. I remember hearing one about someone who called in sick to work one night only to post a picture of herself at a concert that same evening.

Every company now uses social media profiles to perform cursory background checks on potential employees. There are companies that perform this service and there are probably companies out there now who will scrub those profile walls if you need such a service before applying for a job.

◆❯ You are who you are, no matter where you are.

Don't try to project a false image, as time will eventually expose you.

A while back, my wife and I were in the process of buying a house. It was a short-sale and anyone who's gone through this knows how long this process is and how challenging it can be when negotiating with the bank instead of the property owner.

The house was everything we were looking for. In fact, I recently came across a list of things we wanted in a new home and literally this house met twenty of the twenty-five items we had written on that piece of paper two years earlier. The house was in great condition and had enough rooms for our kids in a quiet neighborhood near the school they attended.

After we put in the offer, we waited six months before we were able to close on the property. During this time we went through weeks of no communication with the realtor with moments of rapid hourly communication when the bank was all of a sudden ready to close the deal.

You can imagine the anxiety and frustration with waiting that long and not knowing if we were going to get the house or not. Should we start looking for another home? Can we stay where we are another year while we wait? Questions continued for months without really knowing whether or not this "dream" would come true.

During this time, I ended up "friending" my realtor on Facebook– big mistake.

There were times in the process that we waiting for information from him as we didn't want the deal to fall through.

On one side, he came across as a professional working for one of the larger real estate firms – and on the other side he was sending out immature jokes and posting them to his Facebook page.

I found this contradiction challenging to overcome as I found myself losing respect for him professionally as well as personally.

The reason is that our minds cannot discern the difference between the professional image and the personal image. We only see and comprehend the person as *one being* and over time the messages we receive determine how we perceive that person. This creates a dissonance between who they project themselves to be and how we actually see them.

This dissonance is best explained by the concept of "Balance Theory" developed by Austrian psychologist Fritz Heider.[3] Heider's work studied the relationship between three separate things. In the case of a celebrity product endorsement, the three components are the celebrity, the product, and you. If you like the celebrity, it's likely you'll also like the product.

This is why so many famous actors are used in marketing products *and* why these ads are pulled from the air if/when the famous actor gets into trouble.

If you like the product, but not the celebrity, an imbalance is created and you'll find a way to resolve it. The most common way this occurs is that you'll stop liking the product simply because of the transgressions of the celebrity.

This is why companies across the world have to be extremely careful with whom they associate themselves because it can make or break a company.

You don't have to look far to find some celebrities who were being paid to endorse certain products only to be dropped when their personal lives became in question. Tiger Woods, Lance Armstrong, Paula Deen and Michael Phelps are a few that come to mind.

In a similar fashion, the concept of "the enemy of my enemy is my friend" also proves true in these situations. I may not know someone personally, but if they don't like something that I don't like, then I immediately start to like them.

I went out to dinner with a group of people the other night and the guy next to me ordered a salad and asked the waitress to leave off the mushrooms. Having the same disdain for fungi on my salad, I immediately liked this guy and we were able to strike up a conversation based upon our similar dislike of eating things that grow in cold, dark, moist areas that shouldn't be discussed in polite company.

And for you mushroom lovers out there reading this, you may be experiencing an imbalance as your view of me changes because I don't like mushrooms!

Think of someone you know who always seems to complain about things. These constant gripers drain our energy. You probably don't like spending time with them. They become what they speak about in your mind. It doesn't matter if they are the one who all the bad things are happening to or not, you'll start to associate them with negative feelings.

> **WHAT YOU SPEAK ABOUT BECOMES A BILLBOARD FOR OTHERS TO READ. IF THEY DON'T LIKE WHAT YOU'RE REPRESENTING, THEY'LL STOP SPENDING TIME WITH YOU.**

This becomes important when it comes to hiring the right people. Simply by putting a person on your payroll makes them paid endorsers of the product or service that your company provides. If people don't like them, they'll stop liking your product.

How many times have you had a bad experience with a retail store or a restaurant and said to yourself, "I'll never go there again." This probably had nothing to do with the actual product or service, but rather the person with whom you interacted.

I recently took my family to Breckenridge for the weekend. On our way back home, we stopped in Silverthorne to swim at the recreation center and play at the park before completing our journey home. After a full day of swimming and playing, our kids were starving and had their minds set on some nachos.

As luck would have it, there was a Mexican restaurant right on the way out of town. As we walked up to the counter I asked the guy working the counter for some kids nachos. Without missing a beat, he said we had the wrong place and they don't have any queso. *No queso?!*, I pondered, *What kind of Mexican restaurant is this?* So I responded with something like, "you have chips, grated cheese and a microwave right?" Put it all together and what do you have?

(You should know I too was a little tired and a bit punchy.)

He replied, "Well, I don't have a button on my cash register for that."

At this particular moment in time, I had a choice to make as a look of utter astonishment overtook my face. Thankfully for him, I happened to glance down at my kids and saw their anxious looks awaiting my response. So, I politely asked if there was a different restaurant nearby. He thought there was, but he couldn't tell me where.

As such, we decided to hop in the car and drive across the road and literally with half a mile there was a Mexican restaurant with plenty of queso-laden nachos that were devoured within minutes.

I made a decision right there that I would never eat at that chain again – only because of the experience I had with this employee. Truth is, however, my boycott lasted less than a month. I went back.

Fearless integrity has to be woven throughout every interaction you have with other people. As someone who is frequently on the hiring side of the RFP process, if I sense a slack in personal integrity, you can guarantee that I'll look for someone else to hire. You don't get to pick and choose when you'll have integrity, it just shows up. Once you drop the ball in this area, it's extremely difficult to regain a client's trust.

-

"IT TAKES YEARS TO BUILD A REPUTATION, BUT ONLY FIVE MINUTES TO RUIN IT."
– WARREN BUFFET

-

chapter 7

KAIZEN

THE SEVENTH AND most critical quality of a winning personality is *kaizen,* or *continual improvement.*

In Japanese, the word *kaizen* can be translated to mean "change for the best". After World War II, American occupation forces brought in business experts to help with the rebuilding of the Japanese industry. When used in the business sense and applied to the workplace, *kaizen* refers to activities that continually improve all functions, and involves all employees from the CEO to the assembly line workers.

The Toyota Company is well known for their streamlined processes and how they've implemented systems that encourage all employees to look for ways to continually improve their particular work areas.

It's through these small and incremental changes that make all the difference in making companies more productive and efficient. Some of the key elements employees look for are ways to eliminate overly hard work (*"muri"*) and wasteful processes (*"muda"*).

We all need to look for ways to eliminate the clutter in our lives, both personally and professionally. By strategically looking for those things you will find them. Many of us will go months and years and follow the status quo without ever truly evaluating the things we do to see how they measure up to the way we want things to be.

One of the best ways to do this it to put together a personal mission statement.

Because if you don't know where you're going, then you'll never know when you get there.

By having a personal mission statement you will know how you want to live your life on this Earth and the impact that you want to have. My greatest fear is to just go on living without ever working hard to realize the goals and dreams that I have written down. The greatest regrets in life come not from the things you've done, but from the things you never did. That's why so many people have made a personal bucket list of things to do before they die.

Sometimes amazing and crazy things show up on those lists! Again, people who have passion and vision in their lives are "client-magnets". They work toward achieving the goals they have in life. They often know this because they have a personal mission statement they work toward every year.

So right now, if you have a mission statement written down somewhere, pull it out and read through it. How are you doing? If you don't have a mission statement, stop, pull out a sheet of paper, and answer the following questions to get you started:

1. What do you want to accomplish in the next five years?

2. Who do you want to influence in a positive influence way?

3. What relationships do you want to make stronger (spouse, children, and colleagues)?

4. How do you want to be remembered? What would your obituary say and who would show up to your funeral?

5. What legacy will you leave behind? Will others be better because you lived and specifically how will they be better?

Take this list and write down a couple simple sentences to help you capture how you feel right at this moment. Don't worry about it being perfect or remembering everyone or everything. This is a living document that you'll refine and update over the years.

I just did this exercise again and I can't tell you how exciting it is. We are wired to achieve no matter what that looks like. There's something in the human spirit that desperately wants to do something significant—and continually get better at it. Many people find themselves depressed when they have nothing to look forward to in life. We need an overarching purpose to our lives in order to be healthy.

Here's the exciting thing – your efforts will be compounded in the relationships you have. When you're excited and motivated you'll start to see those closest to you feel excited and motivated as well.

As a boss or supervisor, when your people know and can see that you embrace a program for continual change in your personal life, they will be encouraged to do the same.

**❯❯ STATUS QUO CANNOT BE MAINTAINED.
IF YOU'RE NOT GROWING, YOU'RE DYING.**

Look for ways to develop yourself as you are the one who will receive the greatest benefit. You're also the one who has the most to lose for not taking deliberate action in this regard.

Once you have your mission written down, it will start to activate your subconscious mind to look for ways to accomplish your goals.

If you've ever bought a certain car, you'll start noticing them everywhere. The same is said with your goals. Once you've written them down, your mind will begin seeing people who have achieved those goals—and, more importantly, you'll start seeing opportunities to accomplish them as well.

I was driving down the road the other day listening to some talk-radio show and I was annoyed when a radio personality recommended a book she had recently read. Without going into any specifics about the book, she simply said in a high-pitched, flowery voice, "This book changed my life!"

My initial response to this statement was one of absolute disdain. I can't stand it when people describe one event, or in this case, reading one book, as such an Earth – shattering event that it would be the sole reason that altered her course in life forever. Too many times people take these simple events and claim them as way more than they really are.

I know they are well intentioned, but when someone says this type of thing to you, how do you react? Maybe you feel a little irritated as well. Perhaps this stems from the feeling that, once again, everyone else's ship has come in and you missed it.

The person who tells you how this book has changed his or her life might as well be saying to you, "My life is better than yours. I found the answer, you haven't, don't you wish you were me and so smart and successful, but alas, you hit all the red lights while I fly right through with nothing but green lights for miles?" You can almost see them putting their thumbs to their ears, waving their fingers and smirking while they say, "nanner, nanner, nanner, you can't catch me…"

In any case, after this momentary pity-party, I had to start asking myself the question as to why this statement bothered me so much. Why was it, that in just one phrase spoken over the radio waves, I found myself in the sea of cynics, riding each passing wave of pessimism to such depths of self-pity?

This is when it struck me – I'm asking the wrong questions. My subconscious mind automatically flows to the path of least resistance. That's why it's called subconscious – somewhere below active thinking – therefore, I'm not actively pondering the concepts, I'm just reacting.

As I began to actually think about, but not accept the downward path that I was heading, I had to ask myself a deeper question, which was, "Why did those few words cause such a negative reaction from me?"

Over the next few days I began searching for the answer, focusing on what it truly means to "change your life." After giving this some time to percolate, I came to this conclusion: every experience you have "changes your life" in some way. Sometimes these experiences have huge consequences while others don't even make it to the foot notes on the pages of your life.

This can be likened to a snowflake falling through the sky. Scientists say that no two snowflakes are identical. Each gust of wind, or contact with other flakes, shapes the crystal into a wonderfully unique entity, completely different from all others.

So too are these somewhat insignificant events in our daily lives. But added up over time, they can make all the difference in who we become.

What influences in your life will you allow to direct the path upon which you will travel? It's your choice. You can't always determine what happens to you, but you can decide how you react to it.

Just for today, decide to be changed in a positive way. By either acting or reacting to the cards life deals to you. The hand you hold at the end of the day is partly the cards you are dealt, and partly the cards you decide to discard or keep. It's your choice.

In *The Slight Edge*[1], a fantastic book written by author Jeff Olson, he explains how the difference between successful and unsuccessful people rests in their daily routines. They don't seem like significant actions at the time or even in a year's time, but compounding those actions over a five-year time-frame makes a huge difference.

In his book, Jeff shares a story about a wealthy old man and his twin sons. The old man is close to dying and decides to give his boys a choice on how they will receive their inheritance.

As he explains that he only has but a short time to live, the two boys nearly twenty-one years of age will need to make their own mark in the world. Not wanting them to start empty handed, he pulls out a box that contains a stack of one thousand crisp 1,000 dollar bills – one million dollars in cash. In another box he pulls out a single penny.

He explains that if they choose the one million dollars, they can take the money and do whatever they desire to make more money to secure their future. If instead they choose the penny, he will double the amount left in the box, each day, for the following thirty days.

Both choices were perfectly acceptable and they simply needed to inform him of their decision the following morning.

The first boy stayed up all night trying to calculate which alternative would be the wisest investment. The second boy had already decided to take the cash, so he couldn't wait until morning to begin his journey to wealth.

The next morning, the second boy rushed to his father's side and accepted the cash. He immediately hired a consultant and manager to help him execute his carefully wrought plan. By the end of the first week they hired the best financial advisors in the land. The second week was filled with brainstorming sessions, reviewing proposals seeking the best investment strategies that would turn his million into multi-millions.

By the third week, all the analysis was done and the selected alternative was sure to double his money within the month. Everything was going per his plan so the boy decided to pay his brother a visit since he hadn't seen him since that fateful morning he accepted his inheritance.

When he arrived home he was shocked to hear that his brother took the penny instead of the million dollars. His brother explained that he had been checking in every day on progress of this little penny and just as his father explained they second day there were two pennies; the third day there were four pennies and by the end of the first week he had a meager sixty-four cents.

By the end of the second week when the second brother's financial advisors were narrowing down their investment alternatives, his penny had multiplied to a total of $81.92. Now, just a few days into the third week, his total earnings were up to $655. His brother started to feel sorry for him and scoffed, "I can't believe you took the penny! Hurry, go plead with father before he passes away and maybe he'll allow you to reconsider and possibly get at least half of a million dollars."

The first brother seemed at peace and wouldn't hear of it. Later that week, their father died peacefully in his sleep. As the end of the month was approaching, the second brother's financial advisors were preparing their financial report. The markets, it seemed, had gone a bit soft, taken a bit of a tumble actually. They did what they could to save things and ultimately turned in a nice profit by increasing his million to one and a half million dollars. This was good news, but once the team of experts started deducting their fees and reimbursable expenses that had acquired over the prior month, there was actually just a little under a million dollars remaining.

He quickly went to go visit his brother to see how he fared as the last time they spoke he was sitting on a total net worth of just

$655. This is when the real shock set in. Turns out that the first brother that began with just a simply penny, but doubled on a daily basis had now turned that penny into just over a million dollars by day twenty-eight. By day twenty-nine this amounted to over two million and by day thirty he had accumulated over five million dollars!

This is the beautiful power of compounding interest over time. The actions that you perform on a daily basis may not seem all that consequential, but compounded over time will make all the difference in the world.

This is the basis of the *slight edge* that Jeff Olson brilliantly illustrates in his book.

Do you let another five years pass you by without a second thought? Or do you sit down and make a game plan for the next five years? It's your choice.

The other harsh reality is that some of those who work with you would like it best if you didn't succeed or change in any significant way – for by doing so you simply expose their lack of motivation to change.

Ambition exposes apathy. That may be hard to hear, but it's the truth. Take a good look around you. When someone gets a job promotion, or accepts a higher position with another company, or has some success, watch how people react. Their response can be telling as it's likely the first reaction is one of jealously, not genuine enthusiasm.

How can you take action and start the process of personal change?

Take what you've read today and start applying it to your life. Make a commitment to personal development and stick to it.

If you're having a hard time coming up with some excuses, let you share a few of mine with you:

First, you're too busy. You really don't have the time to commit to doing anything extra-curricular at all.

Second, there isn't anyone else in the organization as focused and disciplined as you, so why bother trying to improve when the bar is already set so low?

Third, you just realized that you're really not all that focused and disciplined after all, so let's just stop right here.

The sad part is that you can actually stop yourself before you've even started. I get that. But here's the thing – if it were easy, everyone would do it.

It's not easy.

Real change often comes dressed in overalls and looks like hard work, simply because it is. No magic formula, no ten-step program, just plain and simple hard work, on a daily basis, to become a little better than you were the day before. That's it.

What happens to a lake with a surface temperature of thirty-three degrees Fahrenheit? Not much, other than you don't want to take a swim. How about at thirty-two degrees? That's right. On a cold winter day, you can actually hear the surface of the water start freezing over like the sound of a small crackling fire. The water begins to harden and expand against the adjacent edges of the lake.

By focusing on your goals and making daily continual improvements (kaizen) in your life, it will seem like the water is slowly freezing as you move toward your goal. Change that comes over time with hard work is lasting. Winning the lottery will not create immediate change in character, it will only expose it. That's why the statistics prove that lottery winners are often broke and in debt within a few years-- because the changes necessary to manage money were not learned and embedded in their character.

Projects will sometimes go bad. But if you've put your effort into becoming better at what you do, no one can take that victory away from you. It's always yours. Think of it as a investing into a sure thing that really is a sure thing. No external or overseas forces can take away the equity you've put into yourself.

As we conclude our time here together in this book, my hope and prayer for you is that you'll institute a strong, personal desire for continual growth. Buy books, e-books, listen to podcasts, read blogs, attend seminars and invest time and money into leadership development programs. With an attitude of a student, you'll always be on the cutting-edge of personal development—and will lead others with a heart of a teacher.

NOTES

chapter 1

1. "Bob Ross," *The Biography Channel website,*
http://www.biography.com/people/bob-ross-9496426 (accessed Jan 03, 2014).

2. John Miller, *QBQ! Question Behind the Question*, Putnam Publishing 2001

3. *Forbes.com*, "People with passion can change the world," Carmine Gallo,
January 17, 2011.

4. William Hague, *"William Wilberforce: The Life of the Great Anti-Slave Trade
Campaigner,"* London: Harper Press (2007)

chapter 2

1. John F. Kennedy speech, Nasa.gov website,
*http://www.nasa.gov/vision/space/features/jfk_speech_text.html (accessed
January 20, 2014)*

chapter 3

1. *CNN Money,* "Internet outraged by Facebook's "creepy" mood experiment," Charles Riley, June 30, 2014.

chapter 4

1. *Mayo Clinic website, http://www.mayoclinic.org/healthy-living/stress-management/in-depth/positive-thinking/art-20043950 (accessed July 5, 2014)*
2. Earl Nightingale, "The Strangest Secret"
3. *ABC News,* "Man documents final adventures with leg before amputation," Gillian Mohney, July 18, 2014

chapter 5

1. *Association for Psychological Science,* "Giving time gives you time," Cassie Mogilner (The Wharton School, University of Pennsylvania), Zoë Chance (Yale School of Management, Yale University), Michael I. Norton (Harvard Business School, Harvard University).
2. *WebMD,* http://www.webmd.com/diet/news/20061103/exercise-fights-fatigue-boosts-energy (accessed July 18, 2014)

chapter 6

1. *"Courageous",* Tri-Star pictures and Sherwood Pictures 2011
2. *ABC News.go.com,* "Fla. Contractor That Fired Lifeguard For Saving Man Outside Zone Backs Down," Alexis Shaw, July 5, 2012.
3. Fritz Heider, *The Psychology of Interpersonal Relations, Psychology Press; 1 edition* (December 1, 1982)

chapter 7

1. Jeff Olson, *The Slight Edge*, Success Books, 2005

PROPOSAL PREPARATION WORKSHEET

1. How will you show the client your passion for this project?

2. What's the client's vision for this project?

3. How will use your IQ (Interesting Questions) to connect with the client?

4. How will use a positive attitude to come across as confident, not arrogant and make your potential client at ease with who they are?

5. What "give away" will you leave the client to show your generosity?

6. What story will you share that exemplifies you and your companies' integrity?

7. How will you show them your desire for continued personal growth?

Take-Aways

Intro

❯ Focus your presentation on your client's needs, wants and desires—not your own.

chapter _1_

❯ Contagious Passion is the first ingredient for the personality that consistently wins projects.

chapter 2

❯ Sell the sizzle, not the steak.

❯ Whatever project you are working on, if you want people to buy into what you're doing, sell the vision, not the details.

chapter 3

▶▶ BE COMFORTABLE WITH WHO YOU ARE AS THIS WILL LET OTHERS BE COMFORTABLE WITH WHO THEY ARE.

▶▶ TO HAVE AN INTERESTING CONVERSATION, ASK INTERESTING QUESTIONS (IQ). THIS IS WHAT WE CALL CONVERSATIONAL IQ.

▶▶ "IF YOU WANT TO BE INTERESTING, BE INTERESTED."

▶▶ FOCUS ON HOW GREAT YOUR CLIENT IS, NOT HOW GREAT YOU ARE.

▶▶ UNPLUG AND CONNECT!

▶▶ IF BUILDING THE RELATIONSHIP IS IMPORTANT TO YOU, THEN PUT DOWN THE PHONE – UNPLUG AND CONNECT.

▶▶ BE SLOW TO JUDGE AND QUICK TO ASSUME THE BEST.

▶▶ WHEN YOU MAKE A MISTAKE – ACKNOWLEDGE IT.

chapter 4

›› HAVING A POSITIVE ATTITUDE ENHANCES CREATIVITY AND ADDS SIGNIFICANT VALUE TO EVERY PROJECT YOU WORK ON.

chapter 5

›› IF YOU FEEL TIME STRAPPED AND OUT OF ENERGY, GIVE SOME TIME AND ENERGY AWAY!

›› ESTABLISHED RELATIONSHIPS ARE THE STRONGEST ASSETS OF ANY TEAM.

›› BE GENEROUS WITH INFORMATION AS IT HELPS BUILD TRUST.

›› IF YOU DON'T GIVE ONE HUNDRED PERCENT TO THE RELATIONSHIP, YOU'LL ALWAYS DOUBT WHETHER THE OTHER PERSON HAS GIVEN YOU THEIRS.

chapter 6

❯ IF YOU WANT OTHERS TO FOLLOW, THEN LEAD. SET THE EXAMPLE AND STANDARD AND WATCH YOUR TEAM FOLLOW SUIT.

❯ MAKE THE RIGHT CHOICE REGARDLESS OF THE CONSEQUENCES.

❯ YOU ARE WHO YOU ARE, NO MATTER WHERE YOU ARE.

❯ WHAT YOU SPEAK ABOUT BECOMES A BILLBOARD FOR OTHERS TO READ. IF THEY DON'T LIKE WHAT YOU'RE REPRESENTING, THEY'LL STOP SPENDING TIME WITH YOU.

chapter 7

❯ STATUS QUO CANNOT BE MAINTAINED. IF YOU'RE NOT GROWING, YOU'RE DYING.

Quotes to Live By

It's better to hang out with people better than you. Pick out associates whose behavior is better than yours, and you'll drift in that direction.

- Warren Buffett, Berkshire Hathaway, CEO

A great attitude does much more than turn on the lights in our worlds; it seems to magically connect us to all sorts of serendipitous opportunities that were somehow absent before the change.

-Earl Nightingale

It's just better to be yourself than to try to be some version of what you think the other person wants.

-Matt Damon

To be yourself in a world that is constantly trying to make you something else is the greatest accomplishment.

-Ralph Waldo Emerson

Nothing in life is more important than the ability to communicate effectively.

- Gerald R. Ford, 38th President of the United States

Listening is the way to gain wisdom because everything you say, you already know.

-John Maxwell

The single biggest problem in communication is the illusion that it has taken place.

-George Bernard Shaw

We act as though comfort and luxury were the chief requirements of life, when all that we need to make us really happy is something to be enthusiastic about.

-Charles Kingsley, Clergyman

THE QUALITY OF A PERSON'S LIFE IS IN DIRECT PROPORTION
TO THEIR COMMITMENT TO EXCELLENCE, REGARDLESS OF
THEIR CHOSEN FIELD OF ENDEAVOR.
-VINCE LOMBARDI, LEGENDARY FOOTBALL COACH

MANY OF LIFE'S FAILURES ARE PEOPLE WHO
DID NOT REALIZE HOW CLOSE THEY WERE TO SUCCESS
WHEN THEY GAVE UP.
-THOMAS A. EDISON

I CAN ACCEPT FAILURE, EVERYONE FAILS AT SOMETHING.
BUT I CAN'T ACCEPT NOT TRYING.
-MICHAEL JORDAN

IT'S FINE TO CELEBRATE SUCCESS BUT IT IS MORE
IMPORTANT TO HEED THE LESSONS OF FAILURE.
-BILL GATES

FAILURE IS NOT FATAL, BUT FAILURE TO CHANGE MIGHT BE.
-JOHN WOODEN

THE GREATEST GLORY IN LIVING LIES NOT IN NEVER FALLING,
BUT IN RISING EVERY TIME WE FALL.
-RALPH WALDO EMERSON

THE TRUTH OF THE MATTER IS THAT YOU ALWAYS KNOW THE
RIGHT THING TO DO. THE HARD PART IS DOING IT.

— NORMAN SCHWARZKOPF

IN A MOMENT OF DECISION
THE BEST THING YOU CAN DO IS THE RIGHT THING.
THE WORST THING YOU CAN DO IS NOTHING.

- THEODORE ROOSEVELT, 26TH PRESIDENT OF THE UNITED STATES

SUCCESS IS ALWAYS TEMPORARY.
WHEN ALL IS SAID AND DONE, THE ONLY THING
YOU'LL HAVE LEFT IS YOUR CHARACTER.

- VINCE GILL, MUSICIAN

IT TAKES 20 YEARS TO BUILD A REPUTATION
AND 5 MINUTES TO RUIN IT.

- WARREN BUFFET

INTEGRITY INCLUDES BUT GOES BEYOND HONESTY.
HONESTY IS TELLING THE TRUTH – IN OTHER WORDS,
CONFORMING OUR WORDS TO REALITY. INTEGRITY IS
CONFORMING OUR REALITY TO OUR WORDS – IN OTHER
WORDS, KEEPING PROMISES AND FULFILLING
EXPECTATIONS. THIS REQUIRES AN INTEGRATED
CHARACTER, A ONENESS, PRIMARILY WITH SELF
BUT ALSO WITH LIFE.

- STEPHEN COVEY

I'VE LEARNED THAT PEOPLE WILL FORGET WHAT YOU SAID,
PEOPLE WILL FORGET WHAT YOU DID, BUT THEY WILL NEVER
FORGET HOW YOU MADE THEM FEEL.

- MAYA ANGELO - POET

CHARACTER IS LIKE A TREE,
AND REPUTATION IS LIKE A SHADOW.
THE SHADOW IS WHAT WE THINK OF IT;
THE TREE IS THE REAL THING.

- ABRAHAM LINCOLN, 16TH PRESIDENT OF THE UNITED STATES

WHOEVER IS CARELESS WITH THE TRUTH IN SMALL MATTERS
CANNOT BE TRUSTED WITH IMPORTANT MATTERS.

-ALBERT EINSTEIN

WISDOM IS KNOWING THE RIGHT PATH TO TAKE ...
INTEGRITY IS TAKING IT.

-MH McKEE

ALWAYS BELIEVE WHAT A PERSON DOES,
NOT WHAT HE SAYS.

-FRED SMITH

**REAL INTEGRITY IS DOING THE RIGHT THING,
KNOWING THAT NOBODY'S GOING TO KNOW
WHETHER YOU DID IT OR NOT.**

–Oprah Winfrey

THE TIME IS ALWAYS RIGHT TO DO WHAT IS RIGHT.

-Martin Luther King, Jr.

**A LEADER IS LIKE A QUARTERBACK.
THEY DON'T GET PAID TO RUN THE BALL.
THEY GET PAID TO PUT THE BALL IN THE RIGHT HANDS.**

- John Maxwell

**WHEN THE BEST LEADER'S WORK IS DONE THE PEOPLE SAY,
'WE DID IT OURSELVES.'**

-Lao Tzu

**LEADERSHIP IS THE ART OF GETTING SOMEONE ELSE TO DO
SOMETHING YOU WANT DONE BECAUSE HE WANTS TO DO IT.**

-Dwight D. Eisenhower

**HE WHO THINKS HE LEADS, BUT HAS NO FOLLOWERS,
IS ONLY TAKING A WALK.**

-John Maxwell

**WHEN TWO MEN IN BUSINESS ALWAYS AGREE,
ONE OF THEM IS UNNECESSARY.**

- EZRA POUND

**NO MATTER HOW BAD SOMEONE HAS IT, THERE ARE OTHERS
WHO HAVE IT WORSE. REMEMBERING THAT MAKES LIFE A
LOT EASIER AND ALLOWS YOU TO TAKE PLEASURE IN THE
BLESSINGS YOU HAVE BEEN GIVEN.**

- LOU HOLTZ

**THE MOST IMPORTANT SINGLE INGREDIENT IN THE FORMULA
OF SUCCESS IS KNOWING HOW TO GET ALONG WITH PEOPLE.**

- THEODORE ROOSEVELT, 26TH U.S. PRESIDENT

**EIGHTY-FIVE PERCENT OF THE REASON YOU GET A JOB,
KEEP THAT JOB, AND MOVE AHEAD IN THAT JOB HAS TO DO
WITH YOUR PEOPLE SKILLS AND PEOPLE KNOWLEDGE.**

- CAVETT ROBERT

**NOBODY CARES HOW MUCH YOU KNOW,
UNTIL THEY KNOW HOW MUCH YOU CARE.**

-THEODORE ROOSEVELT

Good Reads

Blink, by Malcolm Gladwell

Developing the Leader Within, by John Maxwell

Developing the Leaders Around You, by John Maxwell

Good to Great, by Jim Collins

How to Win Friends and Influence People, by Dale Carnegie

Leadership 101, by John Maxwell

Rich Dad, Poor Dad, by Robert T. Kiyosaki

The Greatest Salesman in the World, by Og Mandino

The One Minute Manager, by Ken Blanchard

The 7 habits of Highly Effective People, by Steven Covey

The Slight Edge, by Jeff Olson

Think and Grow Rich, by Napoleon Hill

Who Moved my Cheese, by Spencer Johnson

Winning with People, by John Maxwell

QBQ!, by John Miller

39706365R00075

Made in the USA
San Bernardino, CA
01 October 2016